JOHN GREGORY'S WRITINGS ON MEDICAL ETHICS AND PHILOSOPHY OF MEDICINE

Philosophy and Medicine

VOLUME 57

Editors

CLASSICS OF MEDICAL ETHICS 1

Series Editors

The titles published in this series are listed at the end of this volume.

JOHN GREGORY'S WRITINGS ON MEDICAL ETHICS AND PHILOSOPHY OF MEDICINE

LAURENCE B. MCCULLOUGH

Center for Medical Ethics and Health Policy,
Baylor College of Medicine,
Houston, Texas

KLUWER ACADEMIC PUBLISHERS
DORDRECHT / BOSTON / LONDON

A C.I.P Catalogue record for this book is available from the Library of Congress.

ISBN 978-90-481-5015-1 e-ISBN 978-0-585-32315-2

Published by Kluwer Academic Publishers
PO Box 17, 3300 AA Dordrecht, The Netherlands

Sold and distributed in North, Central and South America
by Kluwer Academic Publishers,
PO Box 358, Accord Station, Hingham, MA 02018-0358, USA

In all other countries, sold and distributed
by Kluwer Academic Publishers,
PO Box 322, 3300 AH Dordrecht, The Netherlands

Printed on acid-free paper

Printed in Great Britain.

TABLE OF CONTENTS

FOREWORD vii

PREFACE ix

PART I

Laurence B. McCullough / INTRODUCTION 1

I. CONTENTS OF THIS BOOK 3

II. A BIOGRAPHICAL SKETCH OF JOHN GREGORY 6

III. GREGORY'S MEDICAL ETHICS 17

A. *Gregory's Problem List: Eighteenth-Century British Medicine* 17

B. *Baconian Experimental Method and Philosophy of Medicine* 24

C. *Hume's Principle of Sympathy* 33

D. *Gregory's Feminine Ethics* 36

IV. GREGORY'S *OBSERVATIONS* AND *LECTURES* 40

V. PRIMARY AND SECONDARY SOURCES RELATED TO GREGORY 44

A. *Primary Sources* 44

B. *Secondary Sources* 49

BIBLIOGRAPHY 53

PART II / TEXTS

MANUSCRIPT SOURCES

Whether the Art of Medicine as its has usually been practiced has contributed to the advancement of mankind. Question 59, Aberdeen Philosophical Society, July 12, 1762. Aberdeen University Library: MS 37, 1762. 59

Lectures on the Practice of Physick, taken by Sir Charles Blagden, 1767-1768. Wellcome Institute for the History of Medicine: MS 2618, 1767. 67

Lectures on the Practice of Physic, reported by J. Foster, 1767-1768. Royal College of Physicians and Surgeons, Glasgow: MS 1/9/5, 1767. 71

Notes taken from Dr Gregory, reporter unknown, 1769. Royal College of Surgeons, Edinburgh: MS D 27, 1769. 85

Notes of Lectures on the Practice of Physic, by H. Morris, 1772. Royal College of Physicians and Surgeons, Glasgow: MS 1/9/10, 1772. 89

PUBLISHED WORK BY JOHN GREGORY

Observations on the Duties and Offices of a Physician, and on the Method of Prosecuting Enquiries in Philosophy. W. Strahan and T. Cadell, London, 1770. 93

Lectures on the Duties and Qualifications of a Physician. W. Strahan and T. Cadell, London, 1772. 161

NOTES 247

INDEX 249

FOREWORD

This volume introduces a new subseries of *Philosophy and Medicine*, *Classics of Medical Ethics*. The purpose of this new subseries is to bring out scholars' editions of major works in the history of medical ethics and philosophy of medicine. This new subseries will target for publication texts that are long out of print and difficult to access.

Each volume will contain an introduction to the writings on medical ethics and philosophy of medicine produced by the original author. Each volume will also contain a guide to the primary and major secondary literature, to facilitate teaching and scholarship in bioethics, philosophy of medicine, and history of medicine. Texts will be presented in their original style and will provide pagination of the original, so that citations can be made either to the original text or to the page numbers in these volumes. Finally, each volume will be well indexed, again to facilitate teaching and research.

Bioethics and philosophy of medicine – the former more so than the latter – have an insufficiently developed understanding of themselves as having a history. As a consequence, these fields lack the maturity that critical dialogue of the past with the present provides for other fields and disciplines of the humanities. To the extent that this problem is due to the fact that major primary historical sources are not readily available, this subseries will contribute to the further development and maturation of bioethics and philosophy of medicine as fields of the humanities.

We look forward to the production of future volumes by scholars and teachers. Volumes in preparation include the works of Thomas Percival and Thomas Gisborne, two central figures of eighteenth-century British medical ethics.

THE EDITORS

PREFACE

The subseries begins with the writings on medical ethics and philosophy of medicine by the Scottish Enlightenment physician and medical ethicist, John Gregory (1724-1773). As the reader will discover in the Introduction to this volume and in the texts themselves, Gregory was a sophisticated student of then contemporary philosophy, particularly David Hume's moral sense philosophy, as well as the Baconian method of observational and experimental science. Gregory's medical ethics also anticipates several aspects of contemporary bioethics, as the reader will also discover.

Support from my colleagues and academic institutions over many years has been essential for the preparation of this book. John McDermott and James Knight at Texas A&M University and Warren Reich at Georgetown University encouraged and supported my interest in and writing on Gregory and other topics in the history of medical ethics. When I came to the Center for Medical Ethics and Health Policy at the Baylor College of Medicine in Houston, Texas, in 1988, Baruch Brody continued this collegial and institutional support. Baruch supported with Center funds a crucial research trip to Scotland and England in 1991, during which I identified three of the manuscript sources that appear in this volume. This research trip was also supported by a Travel Grant from the National Endowment for the Humanities. In addition, Baruch supported my application and found the funding for a sabbatical leave during the 1995-1996 academic year, during which I completed the research for and writing the Introduction for this book. This sabbatical leave was also supported by an American Council of Learned Societies Fellowship that added substantially to my time off for full-time research that year. Additional travel funds for research during my sabbatical year were provided by a Travel Grant from the American Philosophical Society in Philadelphia. During this second trip I identified the other two manuscripts that appear in this volume. This combination of institutional and extramural support made it possible for me to concentrate for a year on my work on Gregory, making much easier the work of the past year of putting the present volume into its final form.

My work, especially on manuscript materials and rare books, was greatly facilitated by truly splendid colleagues on the professional staffs of libraries and rare book and manuscript collections at the Universities of Aberdeen, Edinburgh, and Glasgow, the Royal College of Physicians and the Royal College of Surgeons in Edinburgh, the Royal College of Physicians and Surgeons of Glasgow, the National Library of Scotland, the Royal College of Physicians and the Royal College of Surgeons in London, the John Rylands Library of the University of Manchester in England, the Wellcome Institute for the History of Medicine in London, McGill University in Montreal, the Huntington Library in California, the College of Physicians in Philadelphia, the National Library of Medicine in Bethesda and the Library of Congress in Washington, DC, the Humanities Research Center and Perry-Casteñeda Library at the University of Texas at Austin, the Blocker History of Medicine Collections in the Moody Medical Library of the University of Texas Medical Branch at Galveston, and Rice University's Fondren Library and the Texas Medical Center Library in Houston. Ms. Hannah Glass provided crucial research support at the Osler Library at McGill University in Montreal, Quebec, Canada.

I especially want to thank Colin McLaren and his colleagues, Iain Beavan, Mary Murray, and Myrtle Anderson-Smith for their superb assistance and good cheer while I worked feverishly at the University of Aberdeen on the manuscript materials both of the Gregory Collection and of the Aberdeen Philosophical Society and James Beaton of the Royal College of Physicians and Surgeons in Glasgow for his bringing to my attention materials that play a major role in this book. Michael Barfoot, in a magnificent display of collegiality, put me onto a Gregory manuscript at the Royal College of Surgeons in Edinburgh that, because it was miscatalogued under his son, James', name, I might well have missed. This manuscript appears in this volume. I learned from these colleagues that Texas hospitality is topped by Scottish hospitality. These individuals provide moral exemplars in which their countrymen and countrywomen should take considerable, and justified, pride.

Manuscript material that appears in this book, often for the first time anywhere, is included with the permission of the institutions that own or house them. At the beginning of each manuscript I provide the appropriate acknowledgment. The cooperation and generosity of these institutions was crucial to completing the present volume.

I take immense pleasure in handling and reading manuscripts and old books. At the Library of Congress, for example, I used Thomas Jefferson's copies of works by Gregory. At the Royal College of Surgeons in Edinburgh I had the exquisite pleasure to work with Gregory manuscripts in the magnificent Lister Room at a great leather-inlaid table. I invite the reader to imagine having the same sorts of experiences while reading through and using the texts included here. I have written an Introduction to them, prepared a guide to the primary and major secondary literature, and provided an index to key names, terms, and concepts. Superscripts in the text refer to notes that can be found after the last Gregory text.

INTRODUCTION

John Gregory (1724-1773) made contributions to the history of medical ethics that secure for him the status as a major figure of that history (McCullough, 1998). Gregory wrote the first philosophical, secular, clinical medical ethics in the English language. His work on medical ethics achieved wide influence in Britain, Europe, and America, from the end of the eighteenth century well into the nineteenth century. These achievements alone make Gregory's work on medical ethics worthy of serious attention by scholars in the histories of ideas, of philosophy, of medicine, and of medical ethics and bioethics.

The reader may be struck by the last two items. After all, isn't bioethics – the study, using the disciplines of the humanities, especially philosophy, of ethical concerns in biomedical science, the health care professions, health care institutions, and health policy – a new endeavor, having been invented only in the last three decades? Not really, because Gregory invented philosophical, secular medical ethics in the English language and as we now practice it more than two hundred years later in many countries around the world. We now undertake studies in medical ethics – and bioethics more generally – on the understanding that such studies necessarily combine the humanities, philosophy in particular, and medicine. Gregory wrote his medical ethics in precisely this way, by combining the Scottish moral sense philosophy of David Hume (1978), a philosophy of medicine deeply rooted in Francis Bacon (1875a, 1875b, 1875c), and clinical concerns that ranged from truth-telling and confidentiality, through sexual abuse of female patients and the care of dying patients, to human and animal experimentation. No one before him, in the English language certainly, had done so.

Gregory also forged the ethical concept of a professional, i.e., someone who lives primarily according to fiduciary obligations of service to patients rather than primarily according to the dictates of self-interest. Gregory thus helped to invent medicine as a fiduciary profession, a legacy that persists into our time. Gregory did so in response to the state of disarray in the medicine of his day, a disarray that he meant to set to rights by using the tools of moral philosophy and philosophy of medicine.

Finally, Gregory wrote the first feminine medical ethics in the history of medical ethics. In using this term, I follow Rosemarie Tong, a recognized scholar of feminist and feminine and feminist philosophy, who distinguishes between feminine ethics, based on "gender traits associated with women ... as positive human traits" (Tong, 1993, p. 5), and feminist ethics, which emphasizes political and social agenda to criticize and correct the subordination of women to men (Tong, 1993, p. 6). Gregory wrote his feminine medical ethics on the basis of David Hume's (1978) concept of sympathy, which Gregory gendered feminine, because he thought that women of learning and virtue provided the moral exemplars that physicians – all of whom were then men – should emulate.

Despite Gregory's intrinsic historical importance, his subsequent considerable influence, and his many points of relevance to contemporary medical ethics and bioethics, contemporary practitioners of bioethics and medical ethics do their work largely unaware of his importance for their undertaking and, indeed, largely unaware of the larger history of medical ethics. In my judgment, three main reasons explain this ahistoricism: two flawed but nonetheless influential misreadings of the history of medical ethics; and the lack of contemporary editions for many of the major texts in the history of medical ethics.

The first reason for the ahistorcism of contemporary bioethics and medical ethics concerns the inordinate influence of two deeply flawed views of the history of medical ethics: first, that the history of medical ethics consists in little more than a footnote to the Hippocratic Oath and other ethical texts from the period of the Oath (Baker, 1993a, 1993b, 1995b) and so of little relevance to contemporary ethical concerns that define the field of bioethics; and, second, that writings on medical ethics concern themselves exclusively, or almost so, with matters of etiquette and so are devoid of ethical content (Leake, 1927; Berlant, 1975; Waddington, 1984). The second view has been shown to be mistaken (McCullough 1978, 1983, 1984). More recently Robert Baker (1993a, 1995b) has advanced a decisive critique of both views and shown them to be insupportable. These mistaken views have, nonetheless, resulted in ahistorical bioethics and medical ethics, depriving both of the status as mature fields of the humanities. Recent work in the history of medical ethics has begun to correct this defect in contemporary bioethics and medical ethics (Baker, Porter, and Porter, 1993; Wear, Geyer-Kordesch, and French, 1993; Baker, 1995; Haakonssen, 1997; McCullough, 1998). The purpose the present book is to complement the appearance of these valuable secondary sources with primary source material.

The second reason for the ahistoricism of contemporary bioethics is that when philosophers began to contribute to the supposedly new field of bioethics – theologians (Sperry, 1950; Fletcher, 1954; Ramsey, 1970) led what we must regard historically as the recrudescence of a field, not its invention – in the late 1960's and early 1970's in the United States, the dominant philosophical method at the time, Anglo-American analytic philosophy, was itself quite ahistorical. The result was that, just as the history of philosophy was not regarded as part of the real work of philosophy, so too, history of medical ethics was not regarded to be part of the real work of the exciting new field of bioethics.

The third reason for the ahistoricism of contemporary bioethics and medical ethics concerns the fact that, for most of the major works in the history of medical ethics, no contemporary editions exist. This is certainly the case for Gregory, for whom the most recent edition of his medical ethics lectures dates from an American edition of 1817 (Gregory, 1817) and a British edition of 1820 (Gregory, 1820). As a consequence, only those with access to the few rare book collections with copies of Gregory's published works have the opportunity to read and study his texts. Moreover, relevant manuscript material is even more rare and accessible only to those with the means and time to travel extensively to read and study them in the handful of libraries that hold them among their collections. This lack of a contemporary edition of Gregory's work on medical ethics goes a long way to explain the lack of attention to his work by contemporary scholars and teachers of bioethics. The present volume makes this long inaccessible work available in one place for the first time.

I. CONTENTS OF THIS BOOK

Upon becoming Professor of the Practice of Physic at the University of Edinburgh in 1766 Gregory gave lectures on medical ethics and topics in the philosophy of medicine before his regular series of lectures on the practice of physic or medicine. Medical students at the University of Edinburgh took a series of courses in such areas as chemistry, anatomy, pathology, and physiology in lecture form and then had the opportunity – they had to pay separate tuition for tickets (Risse, 1986) – to experience what we now call teaching rounds at the Royal Infirmary in Edinburgh, which had a ward dedicated to teaching. The faculty staffed this ward, in six-month rotations, following a pattern of teaching and patient care initiated by John Rutherford (1695-1779), Gregory's mentor when he was a medical student at Edinburgh and later his patron for the teaching position he eventually obtained at his alma mater. Gregory thus gave his lectures on medical ethics and the philosophy of medicine just before students had their first academic clinical exposure to patients.

Students took extensive notes in their lectures at Edinburgh, many of which survive for Gregory's lectures on the practice as well as institutions of medicine, as well as for others on the faculty at Edinburgh during the eighteenth century, including William Cullen, John Rutherford, Thomas Young, and others. In addition, students took notes on the clinical cases they saw at the Royal Infirmary. Gunther Risse (1986) has examined these notes in comparison with the published versions of the professors' lectures and with their clinical notes and have found the student notes to be remarkable reliable. We know that Gregory lent his copies of his lectures to his students, giving them an additional opportunity to take down verbatim what he had to say.

Student notes survive from 1767 (two sets) and 1769 (one set) and these are included here. One of these, from the Wellcome Institute for the History of Medicine (WIHM 2618, 1767) contains abbreviated notes, in phrases and sentence fragments, while the other two, from the Royal College of Physicians and Surgeons, Glasgow (RCPSG 1/9/5, 1767), and the Royal College of Surgeons, Edinburgh (RCSE D27, 1769), are written in complete sentences. All three are included here, so that the reader can have a sense of how students received Gregory's lectures and how his thinking developed. These notes record Lecture I of *Observations* and Lectures I and II of *Lectures*. I also include a set of student notes for Gregory's lectures on the history of medicine (RCPSG 1/9/10, 1772), inasmuch as these shed considerable light on his philosophy of medicine, as it was then understood (King, 1978), and concept of the physician as a professional. Gregory addresses the history of medicine in Lecture II of *Observations* and mainly in Lecture IV of *Lectures*.

In 1770 some one or more of his students saw Gregory's preliminary lectures on medical ethics into print, anonymously, under the title, *Observations on the Duties and Offices of a Physician, and on the Method of Prosecuting Enquiries in Philosophy* (Gregory, 1770), which is included here. It was not uncommon at this period for people to first publish their work or have someone do so, anonymously, to test the public's response to it. If the response turned out to unfavorable, then the anonymity of publication, one hoped, could minimize the harm that could come to one's reputation. On the other hand, if the work met with a positive reception, one could then claim it as one's own and follow it with subsequent editions under one's name. The *Observations*, it turned out, enjoyed a positive reception.

Gregory had once before adopted – with considerable success – precisely this strategy, with his first publication. *A Comparative View of the State and Faculties of Man with those of the Animal World* (Gregory, 1765) appeared anonymously in 1765. Lord Lyttleton, a friend from Gregory's days in London in the 1750's played a key role in seeing the first edition of *Comparative View* into print, a book that met with a very positive reception and advanced Gregory's standing as a major intellectual figure of the Scottish Enlightenment. Agnes Grainger Stewart, Gregory's third biographer, describes the impact of this book in the following terms:

> ... the book had been published in London and received such an enthusiasm that even Gregory and his patron [Lord Lyttleton] were greatly astonished. London read the book, Aberdeen read the book, and so did Edinburgh [where Gregory had just moved from Aberdeen and was seeking an appointment at the University], and Gregory was made at once a member of ... literary Edinburgh ... (Stewart, 1901, p. 112)

Subsequent editions of *Comparative View* then appeared under Gregory's name, to continuing acclaim, and played no small role in helping him obtain his appointment to the medical faculty in the University of Edinburgh.

Gregory seems to have followed a similar strategy with his *Observations*. After all, he did lend out his copy of his lectures, so students could take down verbatim notes – a habit that won him a reputation – and affection – among his students for his generosity. Gregory's son James, who would succeed him at the University and go on to a distinguished – and controversial – career in academic medicine, had at the time begun his medical studies at Edinburgh. It may be that James played a role in seeing his father's lectures into print. If one compares the style of the *Observations* with the version that follows them under Gregory's own name two years later as *Lectures on the Duties and Qualifications of a Physician* (Gregory, 1772b), also included here, and with James' own published works, it becomes apparent that the *Observations* has the sharp-edged style of James' work, which John Gregory eschewed in all of his writings. Compare, for example, the vitriolic attack in the "Advertisement" in *Observations* with the very brief "Advertisement" in *Lectures*. However, I have been unable to locate any documentary evidence to link James Gregory to the first publication of his father's lectures on medical ethics and philosophy of medicine. I therefore caution the reader that I offer the suggestion of such a link as speculation, based only on stylistic comparison, a slender reed on which to support such weighty matters (McCullough, 1998).

As we shall see below, Gregory began to work on the ideas that appear in the *Lectures* as early as his medical student days. One crucial period of his intellectual development occurred during the late 1750's and early 1760's when he helped to found and participated actively in the Aberdeen Philosophical Society (Ulman, 1990). In one of his presentations to the Society Gregory first sets out ideas that form the core of his lectures on the philosophy of medicine and so they are included as well (AUL 37, 1758). These re-appear in Lecture II of *Observations* and in Lectures IV-VI of *Lectures*.

II. A BIOGRAPHICAL SKETCH OF JOHN GREGORY

I turn now to the task of providing a biographical sketch of John Gregory, so that the reader will have some sense of the man who produced the works contained in this volume. Alexander Frasier Tytler wrote the first biography of Gregory, to introduce the four volume set of his works (Tytler, 1788). This remains a very useful source, the most complete of the existent biographies. William Smellie (1800) followed with a chapter in a multiple biography and borrows in great measure and often verbatim – writers in the eighteenth century did not share our concern about originality and, therefore, about proper citation and plagiarism – from Tytler's work. Stewart provides a chapter on Gregory in her study, *The Academic Gregories* (Stewart, 1901). This contains some new material vis-à-vis Tytler, but – as was then sometimes the custom – she does not provide citations to the sources from which she worked. Paul Lawrence has provided an indispensable resource on the Gregory family, to which every scholar of John Gregory's work is indebted (Lawrence, 1971). John Truman (1995) has recently provided a brief biography, with useful summaries of Gregory's published work. Christopher Lawrence (1984) sets Gregory in the larger context of the Scottish Enlightenment. There also now exists a detailed, well documented intellectual biography of Gregory in a more current source (McCullough, 1998).

John Gregory was born in June 3, 1724, into one of Scotland's most distinguished academic families (Tytler, 1788; Stewart, 1901; Lawrence, 1971). His forbears included mathematicians and astronomers in the generations beginning with his great-grandmother, Janet Gregorie, and her son, James, Gregory's grandfather, inventor of a type of reflecting telescope. John's father, James Gregorie (1674-1733) – John modernized the spelling of the family name – began the tradition of physicians in the family, undertaking his medical studies in Rheims, Edinburgh, Utrecht, and Paris. James became mediciner or Professor of Medicine at King's College in Aberdeen, which then had no medical school, though it did have one of the first endowed chairs in medicine in Great Britain. John was the youngest of three children that his father had with his second wife, one of the daughters of the Principal Chalmers of King's College.

James Gregorie died in 1733 and Principal Chalmers and John's half-brother (from their father's first marriage), James, took responsibility for rearing John. James succeeded their father as mediciner at King's. Thomas Reid, a cousin, also helped to raise the child. John's formal education began at the Aberdeen Grammar School and then continued at King's College, where he studied Latin, Greek, Ethics, Mathematics, and Natural Philosophy. Reflecting the family "genius," young John did especially well in math. His student notebook from this period displays a keen interest in a wide range of mathematical topics, including "addition, substraction, algebra, plain geometry, heights and distances, and surveying" (AUL 2206/22, 1738).

King's then still retained the Regent system, in which students were taken through the entire curriculum by one professor, in Gregory's case, Thomas Gordon. They formed a close friendship that lasted until Gregory's death. Gordon had an earned reputation for generosity to his students and may have been the exemplar that inspired Gregory to similar treatment of his students during his later academic career.

Gregory followed his father and half-brother into medicine, moving after his studies at King's to Edinburgh, to commence medical studies in 1742. His teachers there included Alexander Monro, *primus* (1697-1767) and John Rutherford. The professors taught largely from the works of Boerhaave (Underwood, 1977). Rutherford helped to initiate the teaching ward at the Royal Infirmary and was among the first to take medical students into the clinical setting for practical instruction. Rutherford also served as Gregory's mentor during this time and later championed him for his academic appointment at Edinburgh.

In 1745, like many Scotsmen and others before him, Gregory moved to Leiden in the Netherlands to continue his medical studies. Although Boerhaave had died, Leiden continued to enjoy an outstanding reputation for its medical studies. In Leiden Gregory took up with other students from Britain, including Alexander Carlyle (1722-1805). Conversation formed a major part of one's education, apparently, exemplifying the liberal arts model of self-education with which we are now familiar. Stewart reports that this conversation was often "brilliant" (Stewart, 1901, p. 103).

> Gregory's great subjects were religion, and the equal, if not superior, talents of women as compared with men. Everybody made fun of him, for 'he could hardly be persuaded to go to church, and there were no women near whom he could wish to flatter;' but he would not change his mind (Stewart, 1901, p. 104).

Three distinct honors came to Gregory during his school days. On April 18, 1743, while he was yet a student in Edinburgh, the City of Old Aberdeen granted him a "Diploma of the Freedom of the City" and elected him honorary burgess of Old Aberdeen (Lawrence, 1971, Vol. 2, p. 406; Vol. I, p. 150). To receive such a high honor at so young an age added, one suspects, to the academic expectations that his family history had already created.

On March 11, 1746, while Gregory enjoyed student life and the company of the English-speaking students in Leiden, King's College awarded him an unsolicited M.D. (Lawrence, 1971, Vol. I, p. 151). Dorothy Johnston (1987) provides a useful account of "irregular" degrees awarded by King's College during this time. Medical degrees were usually granted unearned on the "attestation of two or more doctors, supported by what amounted to a character reference" (Johnston, 1987, p. 138). Gregory's degree was "[s]igned by his half-brother, James Gregory, as Professor and Dean of Medicine, and others" (Lawrence, Vol. 2, p. 406). This would not be the first time that Gregory's career was aided by family connections, a not uncommon phenomenon in these times.

On June 3, 1746, Gregory was elected Professor of Philosophy in King's College, effective upon his return to Aberdeen. John Gregory's school days had come to an end and his days as an academic and practicing physician began. Gregory returned to King's College and took up regenting students. No written materials survive for this period. We can presume that he experienced as a teacher an environment that had not changed too significantly from this student days four years earlier.

Gregory apparently did not finding teaching at King's sufficiently rewarding, because he resigned his position in 1749 and took what we would now call a grand tour of some of the cities of continental Europe. Upon his return he set up in medical practice. He found himself in a market over-supplied with practitioners, including his half-brother, James, who had the added market advantage of being the Professor of Medicine at King's.

Gregory courted and wed, in 1752, Elisabeth Forbes, one of Lord William Forbes' daughters. Tytler describes the Elisabeth Forbes and their marriage this way:

> ... a young lady, who, to the exterior endowments of great beauty and engaging manners, joined a very superior understanding, and an uncommon share of wit. With her he received a handsome addition of fortune [women brought a dowry to marriage]; and during the whole period of their union, which was but for the space of nine years, enjoyed the highest portion of domestic happiness. Of her character it is enough to say, that her husband, in that admired little work, *A Father's Legacy to his Daughters*, the last proof of his affection for them, declares, that "while he endeavors to point out what they should be, he draws but a very faint and imperfect picture of what their mother was" (Tytler, 1788, pp. 33-34).

Stewart reports that "there is a story that her father did not at all approve of the marriage. 'What do you propose to keep her on?' said he [Lord Forbes], and Gregory, getting angry, took his lancet out of his pocket, and said, 'on this' " (Stewart, 1901, p. 107). Gregory, it seems, married the sort of woman whose intellectual abilities and virtues he had extolled during his student days in Leiden.

In 1754, apparently not meeting with the success for which he had hoped upon entering medical practice, Gregory moved to London and "being already known by reputation as a man of genius, he found as easy introduction to many persons of distinction both in the literary and polite world" (Tytler, 1788, p. 34). In London he encountered the world of medical practice in Britain's great urban center, about which more below.

In addition, Gregory also met, and attended the intellectual salons of,

Elizabeth Montagu, known as the Queen of the Bluestockings. The Bluestocking Circle, as it was also called, comprised a group of intellectually talented women whom Mrs. Montagu befriended and gathered around herself, using her own and her husband's considerable wealth to do so. No formal educational opportunities for women existed at this time and so Mrs. Montagu's support was crucial for the intellectual endeavors of some members of the Circle (Myers, 1990). Mrs. Montagu invited the literati of London to her salons so that she and her women friends could have intellectual discussions with men in a context meant to be free of courtship and the frivolity of balls and card games, so that men and women could meet and engage each other as intellectual peers. Mrs. Montagu became Gregory's life-long friend and they exchanged a correspondence in which Gregory developed his feminine ethics and his feminism, thoughts that find their way later in the moral handbook he wrote for his daughters, *A Father's Legacy to His Daughters* (Gregory, 1774).

In 1754 Gregory was made a Fellow of the Royal Society "and, daily advancing in the public esteem, it is not to be doubted, that, had he continued his residence in that metropolis, his professional talents would have found their reward in a very extensive practice" (Tytler, 1788, p. 36). In November of 1755, however, James Gregory died in Aberdeen, "occasioning a vacancy in the Professorship of Physic in King's College Aberdeen" (Tytler, 1788, p. 37). King's offered the Professorship to Gregory and he accepted, returning to Scotland in 1756 "and took upon him the duties of that office to which he had been elected in his absence" (Tytler, 1788, p. 37).

King's College still had no medical school when Gregory returned to Aberdeen and the efforts he undertook with his physician colleague, David Skene, to start one did not meet with success. With his cousin Thomas Reid and others, Gregory founded the Aberdeen Philosophical Society (Ulman, 1990). The Society met at a tavern, the Red Lion, near the campus in Old Aberdeen and members addressed various "questions" and prepared and presented "discourses" on a wide variety of topics in natural philosophy, philosophy, music, and other topics. H. Lewis Ulman has done pioneering work on the Aberdeen Philosophical Society, his book providing an introduction and absolutely crucial guide to the manuscript materials that record the history of the Society. The Scottish Enlightenment was then in full bloom, to which the Society made major contributions, which Ulman describes as follows:

> Thomas Reid's *An Inquiry into the Human Mind, On the Principles of Common Sense* (1764) established the Scottish Philosophy of Common Sense, influenced the shapers of the American republic, and made a lasting contribution to philosophical inquiry. James Beattie's *An Essay on the Nature and Immutability of Truth* (1770) also championed the virtues of "common sense" and earned a wide following in its day, but its reputation has since suffered because it is more polemical and far less rigorous than Reid's work. Alexander Gerard's *Essay on Genius* (1774) is one of the most significant contributions to the "science of human nature" written during the period, and George Campbell's *Philosophy of Rhetoric* (1776) ranks among the most important rhetorical treatises in the Western tradition. Two volumes of somewhat lesser note, John Gregory's *A Comparative View of the State and Faculties of Man with Those of the Animal World* (1765) and James Dunbar's *Essays on the History of Mankind in Rude and Cultivated Ages* (1780), complete the catalog of books that grew directly out of the discourses read to the Philosophical Society (Ulman, 1990, p. 12).

The Society committed itself wholeheartedly to the new Baconian experimental method for natural philosophy, including philosophy of mind and moral philosophy. Gregory's commitment in this regard becomes plain when one reads his "Question 59," "Whether the art of medicine, as it has usually been practised, has contributed to the advantage of mankind" (AUL 37, 1762), which is included in this volume.

The members of the Society also read and debated Hume's *A Treatise of Human Nature*, which had appeared in 1739 (Hume, 1978). *Contra* Hume's later concern, the *Treatise* did not fall stillborn from the press in Aberdeen. The response of members of the Society to the *Treatise* was complex. To be sure, Hume's skepticism did trouble some of the members of the Society, especially when it came to religion or, more properly, natural religion and deism. This reflected what Ulman sees as the earned reputation of the Society, as "representing the conservative, religious opposition to Hume's skeptical philosophy, a sometimes cordial, sometimes contentious confrontation ... " (Ulman, 1990, p. 54). On the other hand, members seem to have taken strongly to Hume's method and the account of sympathy based on that method. Thus, they do not reject Hume's skeptical method in a moderate form, i.e., one consistent with deism, but they do not accept that skepticism when it leads to claims that experimental method excludes God's existence as creator.

Hume's attack on evidence for a creator in the *Treatise* represented, in my judgment, the Humean skepticism they reject, which is decidedly the whole neither of Hume's method nor of its fruits (McCullough, 1998). The intellectual relationship of Society members to Hume therefore exhibited an interesting ambivalence. Gregory's response matches this in every particular, I believe. Gregory accepts the experimental method for natural philosophy and for philosophy of mind. He accepts Hume's major finding from the application of that method to mind and morality, the principle of sympathy. He also accepts a moderate, but not an extreme skepticism (McCullough, 1998).

Hume was aware of the interest of the Aberdeen Philosophical Society in his work through Hugh Blair, a friend of John Farquhar, one of the first elected members of the Society. Farquhar introduced Blair to Campbell and Reid and Blair sent drafts of works by them to Hume for his comments. Blair and Hume exchanged correspondence about the Society's discussion of Hume's work. In 1763 Hume writes to Blair: "I beg my compliments to my friendly adversaries, Dr. Campbell, and Dr. [Alexander] Gerard [both ministers who opposed Hume's irreligiosity], and also to Dr. Gregory, whom I suspect to be of the same disposition, though he has not openly declared himself such" (Burton, 1846, Vol. II, p. 154).

Should Hume should have suspected Gregory to be a possible opponent? Blair passed Hume's remarks on to Reid, and Reid's reply contains clues of the answer to this question:

> Your friendly adversaries, Drs. Campbell and Gerard, as well as Dr. Gregory, return their compliments to you respectfully. A little philosophical society here, of which the three are members, is much indebted to you for its entertainment. Your company would, although we are good Christians, be more acceptable than that of St. Athanasius; and since we cannot have you upon the bench, you are brought oftener than any other man to the bar, accused and defended with great zeal, but without bitterness. If you write no more in morals, politics, or metaphysics, I am afraid we will be at a loss for subjects (Burton, 1846, Vol. II, pp. 154-156).

This letter exhibits a number of interesting features. First, Reid, Gregory's cousin and intimate, does not seem to include Gregory in the category of "friendly adversaries," only Campbell and Gerard. An alternate reading is that all three are *friendly* adversaries, not implacable foes.

Second, Reid expresses debt to Hume for the "entertainment" of the Society, i.e., for offering it a work that required serious intellectual engagement and in this sense of the term the Society was "entertained" in the Questions and Discourses of the Society. Hume's concept of sympathy provided a great deal of such "entertainment" to the Society's members. Third, Hume is both accused *and* defended and, in both cases – especially the first – "without bitterness." In other words, the Aberdeen Philosophical Society took Hume's work seriously, with the diffidence required of intellectually disciplined philosophers.

Fourth, Hume's skepticism regarding religion is not mentioned, an interesting omission. After all, Hume writes in the *Treatise*: "Generally speaking, the errors in religion are dangerous; those in philosophy only ridiculous" (Hume, 1978, p. 272). In this respect Reid's reference to Athanasius is interesting. Athanasius, Church Father and Patriarch of Alexandria, condemned the Arian heresy, which was anti-Trinitarian in its denial that Jesus was one in substance with God the Father. The Athanasian Creed, attributed to Athanasius, was then a lively subject of debate, given the emergence of Arianism, which dissented from the view that Jesus was divine. Reid's comment can be read as the expression of the Society's preference for a skeptic who argues, rather than an unquestioning, unreflective, perhaps enthusiastic believer. 'Enthusiastic' here means a person who does not regulate his or her passions and so they become deformed, because they are allowed to go out of control. On this reading, which is at least plausible, Reid's comment counts as a subtle compliment to Hume's skepticism, perhaps even in some of its applications to religious practice, e.g., to the religious enthusiasm that could then be found in many churches in Scotland and that should not be equated with natural religion.

Gregory led a rich intellectual life during these years in Aberdeen and enjoyed what appears to have been a very happy marriage. The latter ended, as so many did, in loss, with the death of his wife on September 29, 1761, in confinement with her sixth pregnancy (Lawrence, 1971, Vol. I, p. 157, p. 167). The Montagu manuscripts at the Huntington Library include an undated letter, which is most likely from later in 1761, from Gregory to Mrs. Montagu in which he expressed his grief and loss. I find this letter to be completely in the spirit of Gregory the moral sense philosopher and member of the Aberdeen Philosophical Society.

Madam
My Friends used to say that I was Proud of my Wife and my children. I believe I was so. It has now pleased Almighty God to humble that Pride in the Dust. Every One Laments my losing the best of Wives, & my children losing the best of Mothers. But Alas, Madam, I have lost what the World knows not nor can know. I have lost my Friend, my Mistress, the Partner of every Joy & every Sorrow I ever felt since we were United. I was ever sensible of her Superior Genius & Capacity. I felt this but was not hurt by it because she never seemed Conscious of it Herself. I was & I saw with every Sentiment which Love & Gratitude could inspire a Heart of the greatest Sensibility, thoroughly detached from every Pleasure which Life affords & entirely devoted to me & my Children. I perfectly revered her Piety. Religion never appeared in a more amiable form. Most of the time she could spare from the Dutys of her Family & the Dutys of Humanity & Charity were spent in her Private Devotions, & yet she had the most uninterrupted chearfulness & sweetness of temper & the most uniform Vivacity of Spirit I have ever known. These Quality's attached my Heart to Her in the tenderest manner, They engaged my Confidence to so great a Degree that I gave myself & my whole affairs entirely up to her Direction. Our Sentiments, our Tastes, Our Views of Life were the same, We talked of our Childrens Education & all our little matters together, but I left the Execution of Every thing to her, so that for these nine years past I have lived the life of a Child in my own House, & excepting in my own Profession, I gave myself trouble about Nothing. The only Merit I ever took to myself was that my Heart felt & was gratefull for all her Goodness.–The Sensibility of your own Heart will easily represent to you how desolate & forlorn is my present Situation. At a time when I require the greatest Vigor of mind to perform many nameless Dutys which are entirely strange to me I feel my Spirit broke & incapable of any Exertion. The Sight of the Children perfectly unmans me. O, Madam, they were a remarkable Family. This is not the suggestion of Vanity. I think I am now Dead to every Sentiment of Vanity. but indeed they were spoke of all over the Country & considering the attention given to their Education it was no wonder that they were distinguished. Alas I wander. Dear Mrs Montagu, forgive the overflowings of a heart oppressed and feeble. The common Esteem & Affection my Dear Betsy had for you, makes me think my self entitled to pour out my Griefs before you. It is a Melancholy Pleasure, but still it is a very

great Pleasure. I know it has never been in the power of Posterity & an affluent Fortune to harden your Heart against the tender feelings of Humanity. I know it possessed of a Delicate Sensibility, & what is still more uncommon, of a sense of Piety, seldom preserved in the tumult of Public Life, & more generally acquired in retirement or the School of Afliction. Your letter was a new proof of that Goodness which I have so often experienced & shall ever gratefully remember. It suggested the greatest, the only motive I can at present feel for exerting myself, the Consideration of my Infants. This I trust to the Goodness of God, will have its effect, tho the thousand tender melancholy Ideas which every surrounding Object recalls to my Memory, may render it very distant. I always thought cool reason & Philosophy very insufficient Aids to support the Mind under certain Pressures & aflictions. Little did I think how sensible a Proof I was to have (HL MO 1064, n.d., after September, 1761).

Gregory attended his last meeting of the Aberdeen Philosophical Society on November 22, 1764 and moved with his children shortly thereafter to Edinburgh (Ulman, 1990, p. 126). Gregory bought a home on High Street and set up in medical practice. His first book, *Comparative View* appeared and won him access to literary Edinburgh. He was granted Freedom of the City of Edinburgh, a very high honor, on February 12, 1766 (Lawrence, 1971, Vol. II, p. 397). Gregory received his license from the Royal College of Edinburgh in 1765 and was admitted a Fellow of the College in the same year (Royal College of Physicians of Edinburgh, 1882, p. 4).

Before his move to Edinburgh Gregory had sought an appointment in the University there. His wife wrote to Mrs. Montagu in 1760 to ask her to use her husband's influence to obtain Gregory an appointment as Professor of Botany, then a medical subject (Climenson, 1906, Vol. II, p. 226; HL MO 949, 1760). Mrs. Gregory had played the same role earlier, in 1756, while they were in Aberdeen, writing to Mrs. Montagu to see if her husband could help Gregory to obtain an appointment in London at St. George's Hospital, should a vacancy there occur (HL MO 1063, 1756).

Rutherford, Gregory's teacher twenty years earlier, played a key role at this point as his patron. In 1766 Gregory was appointed by the Edinburgh Town Council to be Professor of the Practice of Physic, to succeed Rutherford who had retired, perhaps to make way for his protegé. This caused some controversy, inasmuch as students favored William Cullen, then Professor of Chemistry, for this appointment. When Robert Whytt died later in 1766, the chair in the Theory of Medicine became open. Students published a broadside calling for Gregory to be appointed Professor of Theory of Medicine, Cullen Professor of the Practice of Medicine, and Joseph Black Professor of Chemistry, to succeed Cullen (AUL 404a, 1766). The Town Council was unmoved by the student petition and appointed Cullen to the chair in Theory and Black to that in chemistry.

In 1766, Gregory also gained the appointment to succeed Whytt as First Physician to His Majesty in Scotland (Lawrence, 1971, Vol. II, p. 407). Cullen had apparently vied for this coveted appointment, but lost out. The position paid £100 sterling per annum, a considerable sum.

Students again petitioned the Town Council, suggesting that Gregory and Cullen alternate in the chairs of theory and practice of medicine. The Council agreed to this arrangement in 1769, which continued until Gregory's death four years later.

While in Edinburgh Gregory lectured on the theory of medicine and the practice of medicine. He published *Observations* and *Lectures*, as well as *Elements of the Practice of Physic* (Gregory, 1772a). He also wrote *Legacy*, which appeared posthumously.

In his lectures Gregory addressed the disease that killed him, gout:

> The gout is a disease that is remarkably hereditary and often appears very early in life. Hereditary gout often attacks about twenty or thirty and *this was the case with myself*, but otherwise the gout is a disease that does not come on 'till the middle period of life between forty and fifty (EUL Dc.7.116, 1769-1770. pp. 254-255, emphasis added).

Hereditary gout also causes premature mortality, of which Gregory's students would have been well aware. Gregory's mother had died from hereditary gout (Smellie, 1800, p. 116). Tytler provides this account of Gregory's death:

> These letters [*Legacy*] to his daughters were evidently written under the impression of an early death, which Dr Gregory had reason to apprehend from a constitution subject to the gout, which had begun to shew itself at irregular intervals, even from the 18th year of his age. His mother, from whom he inherited that disease, died suddenly in 1770, while sitting at table. Dr Gregory had prognosticated for himself a similar death; an event which, among his friends, he often talked, but had no apprehension of the nearness of its approach. In the beginning

> of the year 1773, in conversation with his son, the present Dr James Gregory [known for his directness, to the point of rudeness], the latter remarking, that having, for the three preceding years, had no return of a fit, he might make his account with a pretty severe attack at that season, he received the observation with some degree of anger, as he felt himself then in his usual state of health. The prediction, however, was too true; for, having gone to bed on the 9th of February 1773, he was found dead in the morning. His death had been instantaneous, and probably in his sleep; for there was not the smallest discomposure of limb or of feature, – a perfect *Euthanasia* (Tytler, 1788, pp. 78-79).

III. GREGORY'S MEDICAL ETHICS

Gregory wrote his medical ethics in response to intellectual and practical problems in the medicine of his time. He brings to this endeavor his commitments to Baconian experimental method and the philosophy of medicine that Baconian method generates, to Hume's moral sense philosophy and its principle of sympathy as the well-spring of moral judgment and behavior, and to feminine ethics, in which he genders sympathy female and finds in women of learning and virtue the moral exemplars of sympathy rightly developed.

A. Gregory's Problem List: Eighteenth-Century British Medicine

The world of medicine and the sick in London of the period has been described with great scholarly acumen by Dorothy and Roy Porter (1989). Medical theories and their practitioners competed, often fiercely, in the marketplace of ideas and employment. There existed competing accounts of the nature, origin, and management of disease and illness and as many practitioners ready to employ them. No regular pathway into practice existed, although the Royal Colleges did attempt, without success, to exert monopoly control. One could become a physician with or without a degree, or one could arrange for an unearned degree to be granted. Surgeons trained usually by apprenticeship and were organized into their own College or guild. Apothecaries compounded drugs and also diagnosed and treated problems. Female midwives had a flourishing practice, though they found themselves in increasing and intense competition with man-midwives. Man midwifery made its advent with the invention of the forceps and midwifery joined the medical school curriculum. As the Porters rightly put it, eighteenth-century Great Britain experienced an "open world of medicine" (Porter and Porter, 1989, p. 26).

Medical theories had widely ranging levels of intellectual warrant, from none to only slight more than that. As a consequence, treatment was "hit or miss" (Porter and Porter, 1989, p. 27), there was no monopoly of physicians on medical knowledge and practice, no clinical science to evaluate what physicians did, much less the growing concern with documented quality and outcomes-based clinical practice guidelines with which we are becoming increasingly familiar.

Lay concepts of disease enjoyed as much status in the eyes of the sick as medical concepts, or what passed for medical concepts at the time. Moreover, prosperity increased during the century and people had money to spend. The well-to-do sick summoned to their homes physicians, surgeons, apothecaries, midwives – or whomever – as the sick themselves chose.

> Many factors gave pull to the affluent patient. He would do the summoning. He would pay the bill, and so expect to have that say, that sway, which the power of the purse conferred. He might well be of higher social status than the practitioner (Porter and Porter, 1989, pp. 12-13).

There thus arose – Risse (1986) notes this as well – an individual responsibility for illness and a strong sense of self-help (Porter and Porter, 1989, p. 53). Patients routinely diagnosed and treated their own problems; "self-physicking" was a basic survival skill and widely practiced.

> Undergoing medication was not a matter of abandoning oneself blindly to professional authority. It involved active decision-making and negotiation, equivalent to buying an estate or selecting an education for one's children (Porter and Porter, 1989, p. 27).

In such an environment there existed no physician-patient relationship as we know it, i.e., a relationship based on accepted claims by physicians to intellectual authority in the diagnosis and management of disease and illness and the control of the relationship, by and large, by the physician, e.g., in the form of prescriptions and referrals to specialists. The sick could summon the physician or not, get other opinions, choose among them, or even reject them all and "self-physick." Thus, patients "expected to be allowed their say" in the selection of a regimen (Porter and Porter,

1989, p. 87).

The sick selected their physicians on the basis of individual trust: "The trustworthy doctor had tact and a certain pliancy; patients hated feeling bullied, and expected at least to have their wishes respected" (Porter and Porter, 1989, p. 88). The physician's ability to succeed economically depended on his ability to be trustworthy and thus exhibit "a capacity to appeal to the individual and community" (Porter and Porter, 1989, p. 132). The Porters summarize the physician's world, the world of medicine Gregory experienced in London, in the following terms:

> Patient power and disease power between them rendered the physician's lot quite precarious. His capacity to cure was often slight; and he could not count on the state and professional protection that nowadays safeguards the incorporated profession of medicine. Relations between individual practitioners were commonly acrimonious, and intra-professional conflicts flared between the physicians and the apothecaries, between general practitioners, druggists and chemists, and between the regulars and the quacks (Porter and Porter, 1989, p. 117).

A trustworthy character in the physician provided the patient with an anchor in this social storm: "... practitioners did not dictate to clients, lay people were involved in medical decisions, and worry about social *mores* ... and the need for privacy could affect events powerfully" (Wear, 1987, p. 235). As the eighteenth century progressed, however, the confidence of the sick in the ability to identify a physician – or any other medical practitioner – as a person of trustworthy character became problematic. Behavior did not emanate from a vacuum; on this the moral sense philosophers and common sense itself were clear. Behavior emanated from one's person and so gave clues to what sort of person one was, to one's character. Social relations were predicated on this common understanding (Fissell, 1993).

The physician who hoped for success thus learned to "ingratiate himself with polite society" (Porter and Porter, 1989, p. 138). This involved the mastery of the expected manners of a gentleman, in the case of the university-educated physician. Manners went further than (at that time) just how to eat at a dinner table without offending one's fellow diners or (in our time) whether to, and who should, open and hold a door for another. Instead, "... manners both prescribed correct behavior in certain settings and embodied particular ideas about how people lived in groups, how social structures functioned, and how individual conduct and society overall were connected" (Fissell, 1993, p. 19). The key concept here is that people could infer from manners to character, with some degree of confidence.

This assumption was thrown into doubt by the crisis in manners later in the eighteenth century, when "medical manners and morals became unglued; no longer were codes of conduct based on courtesy functional" (Fissell, 1993, p. 32). People mounted attacks on the "insincerity of manners," because of such factors as the "dissimulation" for which the "man of feeling" became infamous and commodification of manners, which could be "bought and sold," depending on whether those manners advanced the interests of the individual who might put them on (Fissell, 1993, pp. 33ff.). In the fiercely competitive world of eighteenth-century British medicine, physicians, obviously, could be counterfeit men of feeling or put on manners as suited their interest in gaining and retaining market share. Fissell neatly sums up the crisis of manners that confronted Gregory and to which Gregory mounts a response:

> ... appearance and reality were uncoupled and courtesy [as well as other manners that might ingratiate one to a potential patient] stripped of moral connotation. Manners could be bought and sold, and so could not function as an indicator of virtue (Fissell, 1993, p. 42).

The ability of a potential patient to select a trustworthy physician, the cornerstone of the physician-patient relationship, becomes problematic in this context. One response involved the creation of "an ethic peculiar to medicine," in which invention Gregory played a central role (Fissell, 1993, p. 42). This crisis of trustworthiness opens up issues of the physician's authority and power, as well as their right use, in the care of patients, a power that obviously is open to abuse (Porter, 1992, p. 103).

The rise of man-midwifery illustrates this crisis. The causes of the increasing involvement of male practitioners in delivery remain something of a scholarly mystery (Loudon, 1993, p. 1051). They appeared as forerunners of obstetricians (Loudon, 1993, p. 1052). This change may have had something to do with the perceived superiority of men in "technical solutions" to medical problems and the introduction of a technology, forceps, in which men were adept (Geyer-Kordesch, 1993, pp. 903-904). The consuming public increasingly preferred the male practitioner to assist delivery (Loudon, 1993, p. 1051), despite the opposition of female midwives (Porter, 1987). Audrey Eccles claims that pregnant patients had rational justification to prefer man-midwives and the advances that they brought to obstetrics (Eccles, 1982, p. 124).

Roy Porter (1987) examines the problems occasioned by the introduction of man-midwives into clinical obstetric management. Some, Porter says, saw the man-midwife "as a sexual infiltrator, a violator of female modesty." Porter quotes from a tract, written by an offended husband, in which such claims are advanced:

> [Man-midwives are allowed] to treat our wives in such a manner, as frequently ends in their destruction, and to have such intercourse with our women, as easily shifts into indecency, from indecency into obscenity, and from obscenity into debauchery (Porter, 1987, p. 217).

The problem that occasioned such an attack was the "manual manipulation of women's private parts," apparently urged by William Smellie (1697-1763), "as integral to [the man-midwife's] allegedly superior techniques of diagnosis and delivery" (Porter, 1987, p. 217).

Philip Thicknesse, Porter reports, attacked Smellie's practices, taking the view that the man-midwife was "a 'sinister figure' " (Porter, 1987, p. 220), a sexual predator aroused by passion induced from touching female anatomy. Obstetrics, Thicknesse argued, should be returned to the province of women, female midwives, where it had flourished for centuries without, among other things apparently, being such a threat to husbands (Porter, 1987, p. 217).

While perhaps at times posing a sexual threat to women – Porter is not able to find much evidence for Thicknesse's accusations – man-midwives did perform useful services, including "colluding in the delivery and subsequent concealment of illegitimate babies" (Porter, 1987, p. 222). Porter summarizes the situation of man-midwives:

> Whether or not there is truth in the allegations made by Thicknesse and others of sexual goings-on between fine ladies and man-midwives, well might the rise of the man-midwife have left husbands feeling dubious about the part they were prepared to play in helping to conceal the fruits of adultery. The enduring role of the doctor as the ambivalent ally of wives in their manoeuvers against their spouses had obviously begun (Porter, 1987, p. 224).

Pregnant women, especially those who were adulterous, may have wondered whether or not a man-midwife, even of ingratiating manners – what better way to gain access to otherwise forbidden bedchambers? – was trustworthy, as would the woman's husband. A "fine lady" seeking to conceal from her husband, family, social circle, and business associates of her husband the birth of an illegitimate child – no small calamity then – needed a man-midwife whom she could trust, who would help to deceive her husband and everyone not present for her confinement with the dissimulation of trustworthiness. A physician, once trusted, might abuse that trust with threats – express or implied – to tell secrets to those from whom the woman had an interest in concealing them. This extreme instability of manners and character illustrates just how deep the crisis of manners could run in the lives of pregnant women and the sick in general, as well as their families.

In the private practice of medicine, then, there existed no physician-patient relationship that could be based on epistemologically authoritative claims about diseases and their management and on moral trust that the physician would act primarily for the benefit of the patient and not his own interest. There existed, instead, a the sick-physician relationship, in which there was no epistemologically authoritative account of diseases and their management and in which the patient, not the physician, held greater power. As a consequence of this power, a physician might find himself summoned to a patient after or before or even simultaneously with competitor physicians, surgeons, apothecaries, midwives, or other practitioners. There were no set rules for how physicians should conduct themselves in such circumstance and so consultation became ethically problematic.

The custom then was for the physician to leave off the care of the dying. Reflecting a long tradition (French, 1993; Nutton, 1993), in his *Medicus Politicus*, a widely influential text, Frederich Hoffmann (1749) had argued that physicians ought to leave off the care of the dying, as matter of prudence. The clergy should attend to incurable cases. If the physician were not so readily to regard cases as incurable and therefore continue to attend the dying, as Gregory argued, clergy-physician relationships also become problematic.

This was also the time when Robert Whytt (1765) was developing the new science of neurology, or the pathology and clinical management of nervous diseases. It was thought at the time that the epidemiology of these diseases involved a higher incidence of them among the well-to-do than among those economically less well off. On the one hand, nervous diseases require a great deal of the physician's attention and patients with them can be difficult to manage, to say the least. On the other hand, such patients require a great deal of attention and therefore many visits and so this diagnosis can produce a handsome revenue stream when made among patients, especially female patients, with the wherewithal to pay for such care.

The Royal Infirmary was a new health care institution, initiated in Scotland in the eighteenth century for the care of the worthy sick poor, those who worked for a living in the emerging businesses of the nascent industrial revolution. Gunther Risse (1986) has written the definitive history of The Royal Infirmary of Edinburgh, which was established in 1739 and by the 1760's, when Gregory taught and practiced there, had grown to more than two hundred beds. This institution was not built for the "community" but for a targeted population, the working poor.

The Infirmary was supported and operated with funds provided annually by patrons. Those seeking admission to the Infirmary had first to obtain a ticket from one of the patrons and then present for admission at the Infirmary. The lay managers of the Infirmary then conducted what amounted to a screening interview, to select against diagnoses with a high mortality, fever (in its various forms) in particular. Thus was born market segmentation, by the first private not-for-profit hospitals in the English-speaking world! Once admitted, the patients were subjected to the strict rules of the institution, enforced by the lay staff. Physicians who attended at the Infirmary did so in an environment in which the lay managers kept a sharp eye on their limited financial resources. Those resources were such that the Royal Infirmary of Edinburgh did not operate at 100% census (Risse, 1986).

Obviously, in such an environment, physicians had more power and control over concepts of diseases and their management. There was thus to be found in the Royal Infirmary the social conditions for the beginning of a physician-patient relationship. The Royal Infirmary of Edinburgh included a teaching ward, which produced a significant share of the Infirmary's income (Risse, 1986, p. 36). To this ward Gregory brought his medical students, who paid for the tickets that would admit them. For many of them, this was their first encounter with people of lower socio-economic classes outside established social hierarchies such as household servants or tradesmen. The teaching ward provided the opportunity for clinical research, especially on pharmacological therapies.

There were charges of abuse raised periodically against physicians who served in the Infirmary, including charges of "cruelty."

> "Gentlemen who arrived to the highest pitch of eminence in their profession use exceedingly harsh language and apparently unfeeling

> treatment to their patients," asserted one author. Another, writing under the pseudonym "Mac Flogg'em," sarcastically suggested to physicians that "you must manifest your authority by being as rigidly severe and contumacious as possible and exert every endeavor to render the situation of the afflicted poor as irksome and miserable as you can." The proper "pomp and Hauteur" could only benefit the institution by driving from the hospital burdensome patients "who had rather leave it half cured than to submit to your insolent and barbarous behavior" (Risse, 1986, pp. 24-25).

Patients came to the Infirmary probably after self-physicking had failed. Thus they brought a combination of "fear and hope," fear that they were gravely ill and hope that admission would save them (Risse, 1986, pp. 182-183). They were thus especially vulnerable to the power of physicians and abuse of that power.

B. Baconian Experimental Method and Philosophy of Medicine.

Gregory's medical studies in Edinburgh and Leiden shaped his thinking in fundamental ways about key topics in philosophy of medicine, as it was then understood (King, 1978). These include scientific method applied to clinical medicine, a concept of the nature of medicine, and the physiologic principle of sympathy. I consider each briefly, in turn.

The reigning scientific method in both medical schools was Boerhaave's, which owed an undeniable debt to Bacon, whom Gregory also read and absorbed completely. Baconian scientific method rejects: (a) dogmatism, an uncritical adherence to authorities; (b) metaphysical methods of science, where 'metaphysical' means the deliverances of pure reason, i.e., *a priori* methods of inquiry divorced from observation and experiment; and (c) systems in the sense of abstract classifications of diseases not based on a careful description of symptoms and clinically-based distinctions of groups of symptoms into disease categories.

In their place Bacon and Baconian scientists who followed after him argue for scientific method based on experiment and observation, which Bacon gathers together under the shorthand, "experience." This empirical method, which we know and practice as the scientific method, became known as the "experimental method."

This method imposes a strict intellectual discipline on the natural philosopher. The investigator of nature establishes "progressive stages of certainty" (Bacon, 1875a, p. 40). The natural philosopher should take due note of these stages, inasmuch as they range from "certainly true," through "doubtful whether true or not," to "certainly not true" (Bacon, 1875c, p. 260). Observation lays the foundation for scientific knowledge:

> ... if in any statement there be anything doubtful or questionable, I would by no means have it surpressed or passed in silence, but plainly and perspicuously set down by way of note or admonition. For I want this primary history to be compiled with a most religious care, as if every particular were stated upon oath; seeing that it is the book of God's works, and (so far as the majesty of heavenly may be compared with the humbleness of earthly things) a kind of second Scripture (Bacon, 1875c, p. 261).

On the basis of observations enjoying the most reliable truth condition, or "certainly true," the natural philosopher proceeds to posit explanations for observed patterns of events that follow each other and associate with very high regularity. The posited explanation, or hypothesis, for Gregory and other Scottish thinkers, takes the form of a discoverable, posited principle, an inherent, constitutive causal force in things that creates the regular association that has been observed. It was more often the case, Gregory notes, that the effects of this principle can be known than its actual mechanism.

The reader might at this point wonder whether this experimental method for natural philosophy involves metaphysical commitments. Indeed, this method does. The "metaphysical" natural philosophy that Baconian natural philosophers reject involves the rejection of alleged *a priori* factors – such as substantial forms – to explain observed causal phenomena. Scientific discoveries, such as the *animalicula* seen by von Leeuwenhoek through his microscope, challenged the static metaphysics of the late Scholastics and Descartes (Wilson, 1995). The new discoveries required causal explanation of regular, patterned change, the regularity of which provided evidence of law-governed change. Friederich Hoffman, for example, proposed an active, animal spirit as such a causal principle that he thought to be really constitutive of the mind (King, 1970).

In other words, the new experimental method did indeed involve a metaphysics, but now of principles of activity, not static form. The static metaphysics of substantial forms was indeed rejected, but not metaphysics altogether. There came into existence the active metaphysics of real, constituent causal principles. Baconians, including Gregory, tended to be scientific realists about causal principles. The new experimental method sought to establish the existence of such causal principles on the basis of "experience," i.e., rigorous observation and careful experiments (usually natural rather than laboratory), and adherence to rigorous rules for hypothesis formation. The new experimental method built on experience and experience alone; hence the rejection of abstract "systems" based on texts rather than direct observation and experiment. The new method was open to anyone willing to submit to its intellectual discipline; hence the suspicion of authorities. Failure to follow the new method in natural philosophy was seen as willful, as a bid for power and authority based on self-interest not the disinterested results of the new method.

Crucially, this new method and the natural philosophy that it produced was thought by most of its adherents to be consistent with deism. Adherents of this method, including Gregory, accepted the argument for the existence of God from design. This is the argument that there is surely an order and stability of nature that we can observe with certainty along with the assumption that such order and stability could not be surd or without explanation. As finite, it could not be the cause of itself and so must have a causal origin outside itself, outside the order of nature, i.e., outside nature itself. Only God fulfills the intellectual requirements of this argument. Thus, secular, experimental method in science, philosophy, and morals is consistent with deism and natural religion.

The experimental method schools its students in what Gregory calls diffidence. This intellectual capacity involves a studied indifference to one's present state of knowledge that requires one to be open to new observations and results of experiments, including especially those that would require one to alter one's present experience-based beliefs. The capacity for diffidence is properly expressed in the virtue of being open to "conviction," i.e., to think anew and even change one's mind in response to new evidence. Thus, adherence to experimental method requires one to be a skeptic in the sense of not accepting any claims of natural philosophy without experience-based support and to questioning accepted belief from whatever source. However, Gregory understood Baconian method not to include the corrosive skepticism of rejecting deism, because the rationality of experimental method presupposes and therefore shows itself to be consistent with deism.

At Edinburgh and in Leiden Gregory encountered Baconian experimental method applied to medicine. Norman Gevitz succinctly summarizes what this involved:

> ... there was also a growing effort to eschew theory arrived at by de-

> duction [reason unaided and uncorrected by experience], and instead to ground medicine upon the direct observation and measurement of phenomena, to conduct controlled experiments and to correlate facts (Gevitz, 1993, p. 605).

Gregory absorbed these lessons whole and they form the basis of his philosophy of medicine.

Gregory puts down his ideas as a medical student, as becomes plain when we consider a key manuscript source, his "A Proposall for a Medicall Society, Written in 1743" (AUL 2206/45, 1743). Gregory emerges in this text as a full-blooded adherent of the new science of medicine. This document begins with a lament for the present state of medicine, despite the "multitude of volumes wrote on the subject" and recent advances in "Anatomy, Chemistry, Botany, & every branch of naturall philosophy connected with the art of heeling" (AUL 2206/45, 1743, p. 1). Gregory's diagnosis goes, he thinks, to the heart of the matter: "The whole plan laid down by L[ord]. Bacon for prosecuting enquirys into nature has been applied in some measure to many branches of naturall philosophy, tho not with that accuracy and fidelity proposed by its great Author" (AUL 2206/45, 1743, p. 1). The same goes for medicine, and this lamentable state of affairs requires remedy, improvement.

> I would therefore propose that a Medicall Society be founded whose proper business may be to presente Medicall Enquirys in the strict method of naturall history & accurate induction from it, & to make an exact separation of those things in the art which are certain & may be depended on, from such as are uncertain or groundless. To make the undertaking more usefull, the Society should prosecute such enquirys only as have a near connexion with the art of heeling or are immediately subservient to the uses of life. In order to this a distinct list should be made out of these Desiderata in medicine q[ch] [which] seem to be of most importance. The Society should appoint a particular member to make a diligent enquiry into one or more of these Desiderata, & report the same to the Society within a limited time (AUL 2206/45, 1743, p. 1).

The rest of this document, considered in detail elsewhere (McCullough, 1998), makes it clear that Gregory had learned very well indeed all of the major lessons of the Baconian experimental method and its chief implication for medicine: medicine must be improved and Baconian method provides the means for doing so.

The "Proposall for a Medicall Society" would come to shape his life's intellectual work. Indeed, under the heading "The Four Capitall Enquirys" (between fifteen desiderata and the general desiderata) he lists the following:

1. The Preservation of Health.
2. The Retardation of Oldage.
3. The Cure of Diseases.
4. The Improvement of our Nature (AUL 2206/45, 1743, p. 6).

The first three express his concept of the capacities of medicine, while the fourth expresses a general philosophy of science. In these elements of a philosophy of medicine, Gregory's exhibits considerable debt to Bacon, who describes the "three offices" of medicine: "the first whereof is the Preservation of Health, the second the Cure of Diseases, and the third the Prolongation of Life" (Bacon, 1875c, p. 383). Bacon elaborates on these "offices," the moral obligations of medicine:

> But this last [prolongation of life] the physicians do not seem to have recognised as the principal part of their art, but to have confounded, ignorantly enough, with the other two. For they imagine that if diseases be repelled before they attack the body, and cured after they have attacked it, prolongation of life necessarily follows. But though there is no doubt of this, yet they have not penetration to see that these two offices pertain only to diseases, and such prolongation of life as is intercepted and cut short by them. But the lengthening of the thread of life itself, and postponement for a time of that death which gradually steals on by the natural dissolution and the decay of age, is a subject which no physician has handled in proportion to its dignity (Bacon, 1875b, p. 383).

The office of prolongation of life "is new, and deficient; and the most noble of all" (Bacon, 1875b, p. 390). As Bacon puts it, "Medicine's there are many for preserving *Health*, and curing diseases, but few to *prolong life* ... (Bacon, 1977, pp. 130-131). The first two offices require improvement, as well. The requirement to improve the three "offices" of medicine Gregory finds still unsatisfied as the writes his "Proposall," as we saw above. Gregory picks this up in the desideratum relating to "History of Old Age with the method of preventing the consequences of it, or the Renovation of youth" (AUL 2206/45, 1743, p. 4), This is geriatrics in its nascent form.

Bacon also addresses subsets of each office. His remarks on those under the preservation of health merit attention. The first concerns incurable diseases:

> Therefore I will not hesitate to set down among the *desiderata* a work on the cure of diseases which are held incurable; that so some physicians of eminence and magnanimity may be stirred to take this work (as far as the nature of things permits) upon them; since the pronouncing these diseases incurable gives a legal sanction as it were to neglect and inattention, and exempts ignorance from discredit (Bacon, 1875b, p. 387).

Gregory has absorbed this point as a medical student and returns to it in his medical ethics lectures.

Bacon also addresses pain management under the rubric of preserving health:

> Again, to go a little further; I esteem it likewise to be clearly the office of a physician, not only to restore health, but also to mitigate the pains and torments of diseases; and not only when such mitigation of pain, as of a dangerous symptom, helps and conduces to recovery; but also when, all hope of recovery being gone, it serves only to make a fair and easy passage from life. ... This part I call the inquiry concerning *outward Euthanasia*, or the easy dying of the body (to distinguish it from that Euthanasia which regards the preparation of the soul); and set it down among the desiderata (Bacon, 1875b, p. 387).

Bacon also faults physicians for not following a regular use of medications, to learn better about their effects.

Gregory undertook his medical studies at a time when the new physiology of the nervous system developed. In the notes that survive from his medical student days Gregory identifies additional topics in a philosophy of medicine, including a vital principle, the mind-body connection, the effect of imagination on the body, and other topics that come directly from this new science. The concept of sympathy plays a major role in this science, because it helps to explain how the anima can affect the body.

Consider general desideratum 5: "A History of the Union betwixt y^{e} [the] mind and body, & their mutuall influence upon one another Particularly of y^{e} [the] power of the imagination not only upon the mind or body of the Imaginant, but upon the mind or body of another" (AUL 2206/45, 1743, p. 6). Imagination reflects the power of the animal spirit, as we saw above, to affect the heart and arouse the passions. Imagination can also affect the mind and body of another person.

In another general desideratum, number 11, he calls for "A History of such things as operate on the body at a distance, the Influence of the Celestial bodys, an Enquiry into what are called Sympethys and Antipathys or such things as operate upon the human body by laws of Nature hitherto unknown" (AUL 2206/45, 1743, p. 7).

One principle of the new science of medicine that obviously caught the young John Gregory's attention at Edinburgh was sympathy. In his "Medical Notes" he includes two headings of interest. The first is the following:

> History of the effects of Sympathy and Antipathy or any other Principles in Nature not generally attributed to, on the Mind and Body (AUL 2206/45, 1743, p. 703).

The rest of page 703 is blank. The second heading appears earlier in the "Notes" and provides us with a clue as to what Gregory was reading at the time. It reads, simply: "Sympathy," under which the following appears: "See practicall remarks on the Sympathy of the part of the body," followed by an abbreviated reference, which turns out to be to an essay by Dr. James Crawford (EUL E.B. 6104 ED 1, 1744). Crawford also studied under Boerhaave in Leiden and taught at Edinburgh for twenty years until his death in 1732 (Underwood, 1977, pp. 99-101).

Crawford's essay concerns "the influence of each part on the other" in the body (EUL E.B. 6104 ED 1, 1744, p. 480). Crawford describes his general view on this topic of the then current physiology of the mind-body complex:

> An accurate knowledge of the Structure and œconomy of the several parts of the human Body, and the influence each particular part hath on another, is the principal Foundation both of Medicine and Surgery ... (EUL E.B. 6104 ED 1, 1744, p. 480).

In reasoning from effects to causes, Crawford identifies three causes for disease in a part, i.e., organ or structure, of the body:

> I. That a Part is affected by *Protopathia* when it is essentially in itself lesed. [diseased], and owes not its Origin to any Communication from another Part. Or by *Idiopathia*, when tho' it be essentially lesed. [diseased], yet the hurt was at first propagated to it from some other Part. Or lastly, by *Sympathy* or Consent, when the Part in itself is yet

> whole and sound, and is only affected by the fault of some other Part ... Diseases by Consent are propagated from a Distance, (in which case only I shall consider them) either by long Muscles or Nerves (EUL E.B. 6104 ED 1, 1744, pp. 482-483).

Sympathy is a physiological principle developed to explain action at a distance – "such things as operate on a body at a distance" (AUL 2206/45, 1743, p. 7). This had posed a deep scientific puzzle for many centuries, in physics as well as in biology (Hesse, 1967). To set Crawford's views – and therefore Gregory's early views – in context, I now turn to a brief history of the development of the concept of sympathy as a physiologic principle.

L.J. Rather explains that in its general sense the concept of sympathy concerns the integrated character of living things, which distinguishes them from non-living entities.

> This is the literal meaning of "sympathy" (*sympatheia*), the term used by the Stoic philosophers to designate a relationship of the kind exemplified by the living body. "Consensus" has a related, if not identical meaning in this context (Rather, 1965, p. 205, n. 2).

In the seventeenth century the concept was further developed to explain action at a distance. Sir Kenhelm Digby reports the case of a man with gangrene. In another room where the patient could not see Digby or what he was doing, Digby placed one of the man's garters that had been in contact with the infection in a solution of vitriol and water. The patient, who did not know what Digby was doing, reported improvement. Digby ordered the patient's plasters removed and then took the garter out of the solution, whereupon the patient worsened. Digby returned the garter to the solution, left it there, and the patient's gangrene healed in six days (King, 1970, pp. 141-142). Digby posited sympathy as the explanatory principle.

The imagination – the capacity to combine ideas that are not combined in experience or reason – can impress itself upon the heart, even though the two are not directly connected. Imagination, by sympathy, can cause a "dilatation of the Heart," arousing particular "Passions" (Digby, 1669, p. 182). There can also be sympathy between a pregnant woman and the infant in her womb, analogous to the vibrating of harp strings when one is struck. Sympathy in this case involves the "harmonious consonance" of atoms in mother and child (Digby, 1669, p. 184).

There can also be "contagion of the imagination" in which the passions of one individual infect others. Melancholy and laughter, for example, can pass from one person to others (Digby, 1669, p. 183). Digby has no explanation for this effect of sympathy.

Hoffmann may have had sympathy in mind as an explanatory principle when he wrote "operations in which concur both the movements of the body and the refection of the mind" (Hoffmann, 1971, p. 11). Jerome Gaub, one of Gregory's teachers at Leiden, provides a somewhat more sophisticated account of the role of sympathy in the mind-body complex:

> The sympathy of the body and mind is ... such that particular affections of the mind will bring on particular disorders of the body, and disorders of the body will in turn affect the mind (Rather, 1965, p. 182).

Crawford picks up on this concept of sympathy and equates it with "consent," which echoes the Stoics on consensus. For Crawford sympathy functions as a principle to explain how a healthy part of the body becomes diseased in a way like another part of the body from which the first body part is at a distance. He posits "long Muscles or Nerves" as the conduit through which sympathy does its work, perhaps following Hoffmann.

Hume, of course, provides an account – the double relation of impressions and ideas – of sympathy as a principle to explain the "contagion" described by Digby, as well as the distress that we experience when we see others in distress. I have been unable to locate any documentary evidence that Gregory had read Hume's *Treatise* while in medical school in Edinburgh. Hume, however, may have been influenced by views such as Crawford's. As a student at the University of Glasgow Hume was member of the Physiological Library (Steuart, 1725). Crawford was, too, and one of his books was included in this library that students and faculty had assembled for their own use. One of Digby's books is listed in the library, as well (Steuart, 1725). Thus, it is very likely that Hume was aware of the history of the physiologic concept of sympathy that I have just adumbrated. Hume's text makes it clear that he treats sympathy as a principle of human psychology, i.e., as a principle of the physiology of the mind that explains the workings of the social principle. Gregory was thus "part of a widespread movement from the middle of the eighteenth century which emphasized the primacy of the nervous system" (Bynum, 1993, p. 346).

C. Hume's Principle of Sympathy

As noted above in the biographical sketch of Gregory, he did indeed read and absorb Hume's moral philosophy in the *Treatise* and its central concept, the principle of sympathy. Sympathy is part of instinct – a strong, active and, when properly developed, inerrant capacity of mind to grasp reality and morality. Sympathy stands in contrast to reason, a weak principle of mind, subject to error when it separates itself from experience, that plays a role secondary and corrective to instinct. Hence, Hume and the other philosophers of the Scottish Enlightenment are known as "moral sense" philosophers.

> No quality of human nature is more remarkable, both in itself and in its consequences, than that propensity we have to sympathize with others, and to receive by communication their inclinations and sentiments, however different from, or even contrary to our own (Hume, 1978, p. 316).

Sympathy "communicates" or acts at a distance, just as the physiological principle described by Digby does. Sympathy causes us to "enter so deep into the opinions and affections of others, whenever we discover them" (Hume, 1978, p. 319). Sympathy is the principle that explains the social principle that binds people together into a national identity. And so we can observe sympathy in children, in people who share national identity, and in myriad other circumstances (Hume, 1978, pp. 316ff).

> We may begin with considering a-new the nature and force of *sympathy*. The minds of all men are similar in their feelings and operations [i.e., there is one mental physiology], nor can any one be actuated by any affection, of which all others are not, in some degree, susceptible (Hume, 1978, pp. 575-576).

How does sympathy work? Hume's answer involves a "double relation" of impressions and ideas. When one sees another human being in pain, for example, on the basis of this impression of suffering one forms an idea of his being in pain. This idea leads naturally and automatically to the impression of pain in one. When properly functioning, i.e., not deformed or defective, as in the case of ingratitude, the double-relation process of sympathy operates automatically. As a result, one has the same impression – or direct, vivid experience – of pain that the individual one observed has.

> This idea is presently converted into an impression, and acquires such a degree of force and vivacity, as to become the very passion itself, and produce an equal emotion, as any original affection (Hume, 1978, p. 317).

Various factors can affect the operation of the principle of sympathy. Sympathy works with perfect strangers. That is, no other individual is a stranger to one. We are all built the same way and function the same way, unless we have fallen into abnormal function or moral deformity, a condition that can be improved. When we resemble others, sympathy will produce impressions of greater vivacity, as will contiguity.

> When I see the *effects* of passion in the voice and gesture of any person, my mind immediately passes from these effects to their causes, and forms such a lively idea of the passion, as is presently converted into the passion itself. In like manner, when I perceive the *causes* of any emotion, my mind is convey'd to the effects, and is actuated with a like emotion. Were I present at any of the more terrible operations of surgery [namely, amputations], 'tis certain, that even before it begun, the preparation of the instruments, the laying of the bandages in order, the heating of the irons, with all the signs of anxiety and concern in the patient and assistants, wou'd have a great effect on my mind, and excite the strongest sentiments of pity and terror. No passion of another discovers itself immediately to the mind. We are only sensible of its causes or effects. From *these* we infer the passion: And consequently *these* give rise to our sympathy (Hume, 1978, p. 576).

Simply being in someone's presence who is in pain powerfully gives one the idea of that individual's pain and automatically an impression of his or her pain. This experience activates sympathy, putting one into action to relieve that individual's pain and thus obliterating all social and other differences, even national identity, that might otherwise make us moral strangers. There can be, in short, no moral strangers in Hume's ethics. Hume on sympathy or humanity – he uses the two words interchangeably – reflects precisely Reid's claim: "Homo sum & nihil humanum a me alienum puto," which Reid takes from Terrence (Reid, 1991, p. 139, p. 316 n. 6). Sympathy moves one into action to respond to both present and future needs of others.

> 'Tis certain, that sympathy is not always limited to the present moment, but that we often feel by communication the pains and pleasures of

> others, which are not in being, and which we only anticipate by the force of imagination. For supposing I saw a person perfectly unknown to me, who, while asleep in the fields, was in danger of being trod under foot by horses, I shou'd immediately run to his assistance; and in this I shou'd be actuated by the same principle of sympathy, which makes me concern'd for the present sorrows of a stranger. The bare mention of this is sufficient. Sympathy being nothing but a lively idea converted into an impression, 'tis evident, that, in considering the future possible or probable condition of any person, we may enter into it with so vivid a conception as to make it our own concern; and by that means be sensible of pains and pleasures, which neither belong to ourselves, nor at the present time have any real existence (Hume, 1978, pp. 385-386).

Sympathy – not reason – governs morality because properly trained and regulated sympathy leads instinctually to one having the same pain or sorrow or distress or pleasure as another and being moved by this impression to appropriate action. This is not empathy, wherein one has an idea of another's plight, links it to oneself in some imaginative fashion, and then suffers for or with that unfortunate individual (McCullough, 1998). Humean sympathy causes one to suffer *just as another suffers*, to find that unpleasant – to be made uneasy Hume says in many places – and moved to action to resolve the uneasiness.

Sympathy creates an unease that can lead to both negative and positive sympathetic responses to a situation.

> I have mention'd two different causes, from which a transition of passion may arise, *viz*. a double relation of ideas and impressions, and what is similar to it, a conformity in the tendency and direction of any two desires, which arise from different principles. Now I assert, that when a sympathy with uneasiness is weak, it produces hatred or contempt by the former cause; when strong it produces love or tenderness by the latter (Hume, 1978, p. 385).

In Hume's *Treatise* sympathy is a moral, mental, instinctual, physiologic principle of human nature that explains the "contagion" of behavior and passions (Hume, 1978, p. 317). More to the point, sympathy explains the natural, built-in, evident, observable, and fundamental other-regardingness of human beings. Sympathy is a real principle, a constituent element of human nature, distinct yet inseparable from the other principles of human nature and from the corpuscles, as it were, of human nature, impressions and ideas through which sympathy acts. Hume summarizes:

> Thus it appears, *that* sympathy is a very powerful principle in human nature, *that* it has a great influence on our taste of beauty, and *that* it produces our sentiment of morals in all the artificial virtues. From thence we may presume, that it also gives rise to many of the other virtues; and that qualities acquire our approbation, because of their tendency to the good of mankind (Hume, 1978, pp. 577-578).

In summary, Hume takes a thoroughgoing, Baconian approach to the science of man with a particular view to generating a science of morals. Hume thus makes good on the promise of the *Treatise's* subtitle: "Being an Attempt to Introduce the Experimental Method of Reasoning into Moral Subjects" (Hume, 1978). Hume picks up on the main themes of the new physiology, with which he was familiar since his student days at the University of Glasgow and his membership in the "Physiological Library" (Steuart, 1725). With the new physiologists, Hume holds that the mind is an active power and sympathy one of its main causal principles.

D. Gregory's Feminine Ethics

Gregory bases his medical ethics on Hume's principle of sympathy, which he had studied and accepted while in Aberdeen, while participating in the intellectual life of the Aberdeen Philosophical Society. Sympathy that moves one to respond positively to another grounds the virtue of tenderness. In many places Hume himself refers to sympathy as "tender." Eighteenth-century British writers used 'tenderness' to denote a female virtue. Hume, as I read him, genders positive sympathy female, as does Gregory. Gregory understands sympathy, properly developed, to generate tenderness: an asexual virtue that moves us to enter into the suffering and distress of others appropriately, with women of learning and virtue the exemplars that show us what 'appropriately' means.

Gregory's views on women can be found in the letter he wrote to Mrs. Montagu when his wife died, quoted in full above. Consider, first, what he writes about his now deceased wife. Gregory pens a portrait of a woman whom he not only loved, but whom he plainly admired for her intellectual capacities and moral character. They had for their short nine years together a marriage he describes with the most admired terms of the day: "Friend, Mistress, and Partner," an equal. Elisabeth Gregory also possessed not just intellectual ability, but "*Superior* Genius and Capacity" (emphasis added), not unlike another "academic" Gregory, his great-grandmother Janet. Elisabeth had cultivated the best sentiments of heart, as well as piety, and "chearfulness & sweetness of temper & the most uniform Vivacity of Spirit I have ever known." For Gregory, female virtues of such exquisite development occupied the highest plane, so that his confidence in his wife was justifiably complete. His family was "remarkable," he says in effect, because of her unflagging efforts, including the education of their children, girls and boys alike, it seems.

In this letter Gregory goes on to put Mrs. Montagu on the same plane with his wife. Mrs. Montagu lives a life unlike that lived by Mrs. Gregory, namely a "public life" in her salons and writings. Her heart had not been hardened "against the tender feelings of Humanity" by city life and the high-powered intellectual circles in which she moved. She is an exemplar of tenderness, the very word Hume uses to characterize sympathy, a term interchangeable with 'humanity'. Mrs. Montagu, too had suffered great loss with the death in his infancy of her only child. Finally, Mrs. Montagu and Mrs. Gregory were distantly related. One of her kinfolk had died. I cannot find the letter from Mrs. Montagu's to Gregory to which he makes reference. Presumably Mrs. Montagu wrote first, upon learning of the death of Gregory's wife. She exemplifies the same "Goodness" as did Mrs. Gregory, the very goodness of heart , i.e., character, that well-developed sympathy should produce as one of its main moral effects. For Gregory, Mrs. Montagu and Elisabeth Gregory were women of learning and virtue, exemplifying properly regulated passion. Properly regulated passion, even though its exemplars are female, does not unman because regulated passion involves the properly cultivated expression of passion and thus human virtue. Only vice or weakness can unman someone. These themes appear in Gregory's treatment of sympathy, as we shall see just below.

The correspondence at the Huntington Library of letters from Gregory to Mrs. Montagu from 1766 until just before his death in 1773 contains many passages in which Gregory expresses his view that women of learning and virtue provide the moral exemplars of properly cultivated and functioning sympathy. For example, he writes to Mrs. Montagu on October 11, 1766:

> There is a nobleness in your Mind which sets you above the Ceremony & the Consciousness You must have the Benignity of your own Heart

should naturally make it open & unreserved (HL MO 1066, 1766, p. 2).

Now, Gregory advanced his femine ethics at a time when it was at risk of meeting with a hostile reception, and it did. We saw above in the biographical sketch that his views on women drew negative responses early in his life, from his fellow students in Leiden. He also presented these views at the Poker Club, one of Edinburgh's most prestigious intellectual societies, and they met with laughter. Gregory remained undeterred.

Gregory – no surprise – goes to some lengths to defend his feminine ethics. Consider the following passage from his *Lectures*, in which Gregory presents his core views on sympathy:

> I came now to mention the moral qualities peculiarly required in the character of a physician. The chief of these is humanity; that sensibility of heart which makes us feel for the distresses of our fellow creatures, and which, of consequence, incites us in the most powerful manner to relieve them. Sympathy produces an anxious attention to a thousand little circumstances that may tend to relieve the patient; an attention which money can never purchase: hence the inexpressible comfort of having a friend for a physician. Sympathy naturally engages the affection and confidence of a patient, which, in many cases, is of the utmost consequence to his recovery. If the physician possesses gentleness of manners, and a compassionate heart, and what Shakespeare so emphatically calls "the milk of human kindness," the patient feels his approach like that of a guardian angel ministering to his relief: while every visit of a physician who is unfeeling, and rough in his manners, makes his heart sink within him, as at the presence of one, who comes to pronounce his doom. Men of the most compassionate tempers, by being daily conversant with scenes of distress, acquire in process of time that composure and firmness of mind so necessary in the practice of physick. They can feel whatever is amiable in pity, without suffering it to enervate or unman them. Such physicians as are callous to sentiments of humanity, treat this sympathy with ridicule, and represent it either as hypocrisy, or as the indication of a feeble mind. That sympathy is often affected, I am afraid is true. But this affectation may be easily seen through. Real sympathy is never ostentatious; on the contrary, it rather strives to conceal itself (Gregory, 1772b, pp. 19-20).

The first sentence describes sympathy as Hume would. Gregory then uses the striking female metaphor of "milk of human kindness" from Shakespeare's Macbeth, to describe sympathy. Lady Macbeth, early in Act I, urges her husband to desire the power for which she had ambition and for which, it seems, he lacks ambition.

Glamis thou art, and Cawdor, and shalt be
What thou art promised. Yet do I fear thy nature;
It is too full o' th' milk of human kindness
To catch the nearest way. Thou wouldst be great,
Art not without ambition, but without
The illness should attend it (Shakespeare, 1982, p. 50).

The editor explains 'milk of human kindness' as the "gentle quality of human nature" and 'illness' as "wickedness," or the relentless pursuit of self-interest (Shakespeare, 1982, p. 50, notes). In other words, Macbeth lacks the relentless ambition of unqualified self-interest. Instead, his ambition is limited (rightly for Gregory, wrongly for Lady Macbeth) by concern for and obligations to others, the milk of human kindness. As we know, Macbeth does indeed become ambitious *with* illness – unbridled self-interest and utter lack of concern for and obligations to others – and thus destroys himself and, nearly, Scotland. Gregory's students would know the passage; Shakespeare was undergoing a renaissance at the time.

The regular exercise of sympathy in response to the suffering of patients schools the physician in steadiness, a regulated response in sympathy expressed as "composure and firmness of mind." This is what Mrs. Montagu displayed in her life, with the loss of her infant son (Myers, 1990). Gregory displays it in the letter he writes to Mrs. Montagu when his wife dies.

True sympathy does not either "unman" or "ennervate" – both seem to be sexual metaphors, the latter particularly striking in this respect – the physician. One need not fear becoming a dissipated ingenue, lounging upon a couch. Nor need one fear the ridicule of colleagues who would say one had a "feeble" mind, a negative, sexist appeal. Just the opposite. Real sympathy reveals strength of character and thus "strives to conceal itself." In this way the physician achieved "benignity" of heart, in which he is "open and unreserved" to the suffering of his patients, just as Gregory describes Mrs. Montagu.

IV. GREGORY'S OBSERVATIONS AND LECTURES

Gregory, as the reader will discover, was well aware of the problems in medicine described above, because he addresses all of them in *Observations* and *Lectures*. He begins *Observations* and *Lectures* with a definition of medicine in terms of its capacities, taken whole from Bacon. He then frames his general ethical concern in terms of the dichotomy between medicine as a trade, pursued for self-interest, and as an art, pursued for the interest of medical science for the benefit of patients. This concern already occupied him in Aberdeen, as is plain from Question 59 that he addressed for the Aberdeen Philosophical Society (AUL 37, 1762). For Gregory, as a moral sense philosopher committed to Baconian method and to Hume's principle of sympathy, the pursuit of self-interest – he calls it, simply, "interest" – represents anathema. In science, the pursuit of interest corrupts medicine, as is plain when he discussed experimentation. There he expresses the fear that rivals in the market place will judge one's innovation negatively simply to preserve their power and market share. The intellectual virtue of the capacity for diffidence, namely, openness to conviction – i.e., change in one's beliefs – shapes the moral core of the Baconian natural philosopher. In clinical practice, the pursuit of interest hardens the physician's heart against human suffering, while over-responding to the sufferings of patients leads to the moral wasteland of dissipation.

Gregory then sets out and defends sympathy as the physician's chief virtue or quality – one of the "qualifications" of the title for the work – as we saw above. The moral virtues of tenderness and steadiness, in which properly cultivated and functioning sympathy is expressed, shape the moral core of the Humean clinician. There thus exists a profound synergy between Gregory's philosophy of medicine and his moral philosophy. This is why he must and does address both in *Observations* and *Lectures*. Together his philosophy of medicine and his moral philosophy shape the foundations of his medical ethics.

Gregory then deploys sympathy and its virtues of tenderness and steadiness in analyses and arguments concerning a wide range of topics. These sections of *Observations* and *Lectures* can usefully be read as a response to the "problem list" discussed above. Gregory addresses conflicts of interest, the governance of the patient by the physician, the care of patients with "nervous ailments," changes in practice style as the physician ages, confidentiality – especially concerning female patients, sexual abuse of female patients (only in *Observations*), temperance and sobriety, laying medicine open (reflecting his commitment to diffidence and its virtue of openness to conviction), truth-telling (particularly in the case of grave illness), abandonment of dying patients, cooperation with clergy, consultation (which does *not* involve etiquette or the mutual pursuit of self-interest), relationships between younger and older physicians (reflecting the problem of intense market-place competition), regard for older writers and medical writings, the boundaries between medicine and surgery and between medicine and pharmacy (which were hotly contested, indeed), formality of dress (again, as the reader will discover, not entirely a matter of etiquette), singular manners (addressed to the problem of the man of put-on, purchased, false manners), avoiding a reaction of disgust to unpleasant clinical situations, time management, servility to one's social superiors who are patients, secrets and nostrums, disclosure of the composition of secret remedies and nostrums to patients (in a treatment far different from our understanding of informed consent), the physician's responsibility when patients die, medicine and religion, experiments on patients, animal experimentation, and the obligations of professors of medicine. He addresses these topics in Lecture I of *Observations* and Lectures I and II of *Lectures*; the last three are addressed in later sections of *Observations* and *Lectures*.

In his lectures on the institutions of medicine Gregory touches briefly on the concept and clinical determination of death. Since we have no reliable such concept, our clinical criteria should be the most conservative, he argues. Gregory wrote at a time when fear of premature burial – not premature transplant of unmatched organs – concerned many people.

> In short in what Death consists we cannot say, because we do not know in what life consists. Haller says that Death takes place when the Heart ceases to be irritable: But we do not know in what the irritability of the heart consists. If a person has all the appearances of death where the vis vitae has been gradually extinguished in consequence of any disease & we have not doubt of burying him. But putrification is the only mark that should determine sudden Death, and this particular happens in consequence of nervous disorders where they have had all the marks of apparent death ... (EUL 2106 D, 1773, pp. 168-169).

In an era in which people lived in fear of being buried alive, such caution seems justified.

Despite the brevity of this treatment, putrefaction does meet the criteria for an adequate determination of death, as set out by the President's Commission for the Study of Ethical Problems in Medicine and Biomedical and Behavioral Research (1981). Very few putrefying corpses will be the living bodies of individuals and there will be very few classifications of dead bodies as alive. The determination can be made without "unreasonable delay," especially of funeral rites. This clinical criterion applies to all clinical situations and is "explicit" and "accessible to verification" (President's Commission, 1981, p. 161).

As the reader can see, Gregory anticipates much of the contemporary agenda of bioethics. As the reader turns now to the texts themselves, he or she will encounter a leading thinker of the Scottish Enlightenment, who wrote a medical ethics that is at once professional, secular, philosophical, feminine, and clinical (McCullough, 1998). He did what we in bioethics now do, two centuries before we thought of doing it.

In response to a world of medical practice marked, if not defined, by the pursuit of self-interest, he argues in his medical ethics and philosophy of medicine that the physician's behavior should be based on properly formed, i.e., systematically other-regarding, intellectual and moral character. His secular, philosophical method leads him to identify both a systematically other-regarding, self-effacing, self-sacrificing intellectual capacity of human nature, diffidence, and its intellectual virtue of openness to conviction and a systematically other-regarding, self-effacing, self-sacrificing moral capacity of human nature, sympathy, and its moral virtues of tenderness and steadiness. These twin capacities and their attendant virtues work in synergy to form the intellectual and moral character of the physician as the fiduciary of the patient. The physician can claim to know with intellectual authority what is in the patient's interest. The physician's own self-interest is systematically blunted, so that he focuses on the patient's interests as his primary consideration. The physician is moved, as a matter of habit, to protect and promote the patient's interests, accepting self-sacrifice whenever required to do so. This ethical concept of the physician as professional has a substance and content lacking in some sociological concepts of professionalism (Berlant, 1975; Waddington, 1984).

In forging this ethical concept of the physician as a true "professional," a word then in currency, and thus giving 'professional' substantive philosophical content, Gregory breaks decisively from the Royal Colleges. Gregory also breaks from Hoffmann, who was explicitly theological and whose admonitions include a strong dose of self-interest. Further, Gregory breaks from Boerhaave and his own teachers and colleagues at the University of Edinburgh by providing a thoroughgoing philosophical basis for the duties and qualifications of a physician. Gregory therefore can be credited with inventing secular, philosophical, professional medical ethics in the English-language literature of the history of medicine. He thus wrote a medical ethics for the physician-patient relationship, grounded in the intellectual and moral virtues of the physician.

Gregory also writes the first feminine medical ethics, certainly in the English-language literature and perhaps in the history of medicine as a whole. He genders sympathy and its virtues feminine and argues that the physician-patient relationship should be asexual and disinterested (i.e., not based on self-interest), though *not* detached. The physician should, when it comes to matters of self-interest, be systematically self-effacing and self-sacrificing. Women of learning and virtue provide the role models for physicians, who should, in response, be gentlemen of gallantry and honor, just as men were expected to be at Bluestocking gatherings and just as they should be in courting his daughters.

Finally, Gregory's medical ethics is clinical through and through. His abhorrence of "systems" and their abstractions keeps him firmly rooted in clinical experience, a leitmotiv of his thought that takes root in his medical school writings and culminates in *Observations* and *Lectures*. For Gregory medical ethics was necessarily clinical, making ethical theory and its clinical application seamless. The phrase, 'applied ethics', we can now say, is just the artifact of "systems" and should not be used to understand Gregory's accomplishments: The distinction between "theoretical" and "applied" is simply foreign to how Gregory does medical ethics.

Gregory's original and formative contribution to the history of medical ethics and therefore to the history of medicine was to combine Baconian openness to conviction and Humean sympathy *to provide medicine with an intellectual and moral identity* that it had until then lacked, at least in Britain and the English-speaking world, and to do so in a secular fashion that both religious and nonreligious individuals would have to accept because openness to conviction to the results of experimental method requires them to accept it. In other words, Gregory writes the first English-language, philosophical, secular ethics of the physician-patient relationship as a truly professional relationship. Before Gregory this literature comprised advice to the "good and learned physician" and admonitions based on Christian ethics (Wear, 1993). Gregory thus provides for medicine the intellectual and moral identity that he saw that it lacked.

V. PRIMARY AND SECONDARY SOURCES RELATED TO GREGORY

To aid those interested in further research on Gregory's contributions to medical ethics, philosophy of medicine, philosophy of science, philosophy of religion, the theory and practice of medicine, and women's studies, the following guide to primary and secondary resources may be of some use.

A. Primary Sources

Publications of John Gregory:

Gregory, J.: 1765, A Comparative View of the State and Faculties of Man with those of the Animal World, J. Dodsley, London.

Gregory, J.: 1770, Observations on the Duties and Offices of a Physician, and on the Method of Prosecuting Enquiries in Philosophy, W. Strahan and T. Cadell, London. Reprinted in the present volume, pp. 93-159.

Gregory, J.: 1772, A Comparative View of the State and Faculties of Man with Those of the Animal World, 5th. ed., J. Dodsley, London.

Gregory, J.: 1772, Elements of the Practice of Physic: For the Use of Students, Balfour and Smellie, Edinburgh.

Gregory, J.: 1772, Lectures on the Duties and Qualifications of a Physician, W. Strahan and T. Cadell, London. Reprinted in the present volume, pp. 161-245.

Gregory, J.: 1774, A Father's Legacy to His Daughters

Gregory, J.: 1779, A Father's Legacy to his Daughters, by the late Dr. Gregory, A New Edition, T. Strahan, W. Cadell, and R. Creech, Edinburgh. (The visitor to the Library of Congress can partake of the special pleasure of reading from the copy owned by Thomas Jefferson.)

Gregory, J.: 1788, The Works of the Late John Gregory, M.D., A. Strahan and T. Cadell, London, W. Creech, Edinburgh. Includes : A. F. Tytler (Lord Woodhouselee), 'An account of the life and writings of Dr John Gregory', as well as Gregory's Comparative View, Elements of the Practice of Physic, Legacy, and the Lectures.

Gregory, J.: 1805, Lectures on the Duties and Qualifications of a Physician. Revised and Corrected by James Gregory, M.D., W. Creech, Edinburgh, and T. Cadell and W. Davies, London.

Gregory, J.: 1817, Lectures on the Duties and Qualifications of a Physician, M. Carey and Son, Philadelphia, Pennsylvania.

Gregory, J.: 1820, On the Duties and Qualifications of a Physician, New Edition, J. Anderson, London.

Translations of Lectures:

Gregory, J.: 1778, anonymous (trans.), Vorlesungen über die Pflichten und Eigenschaften eines

Arztes. Aus dem Englischen nach der neuen und verbesserten Ausgabe, Caspar Fritsch, Leipzig.
Gregory, J.: 1787, B. Verlac (trans.), Discours sur les devoirs, les qualités et les connaissances du médicin, avec un cour d'études, Crapart & Briands, Paris.
Gregory, J.: 1789, F.F. Padovano (trans.), Lexioni Sopra i Doveri e la Qualita di un medico, Gaetano Cambiagi, Florence.

Manuscript and Other Unpublished Materials: Each manuscript source is identified by a multi-letter, capitalized abbreviation, which indicates the library holding the manuscript. The code for abbreviations is the following:

AUL = Aberdeen University Library
CPP = College of Physicians, Philadelphia
EUL = Edinburgh University Library
HL = The Huntington Library, San Marino, California
NLS = National Library of Scotland, Edinburgh
RCPE = Royal College of Physicians of Edinburgh
RCPSG = Royal College of Physicians and Surgeons of Glasgow
RCSE = Royal College of Physicians of Edinburgh
WIHM = Wellcome Institute for the History of Medicine, London

Manuscripts are listed alphabetically by abbreviation. Additional manuscripts of student notes of the kind listed below can be found in the *Biblioteca Osleriana* at McGill University in Montreal, Quebec, Canada, and at the University of Kansas Medical Center Library in Kansas City, Kansas.

Aberdeen University Library
Note: The information regarding titles and dates of material for the Aberdeen Philosophical Society is based on the information in Ulman, 1990.

AUL 37: 1762, 'Dr Gregory – Whether the art of medicine, as it has been usually practised, has contributed to the advantage of mankind', July 12, 1761, Question 59 for the Aberdeen Philosophical Society. Included in the present volume, pp. 59-66.
AUL 129: 1769 & 1770, 'The clinical lectures of Dr Gregory in the Royal Infirmary at Edinburgh Annis 1769 & 1770', 'Taken by G. French'.
AUL G404a: 1766, 'Address of the students of medicine to the Right Hon. Lord Provost, Magistrates, and Town-Council of the City of Edinburgh'. Printed broadside.
AUL 2206/22: 1738, 'Notebook' of 1738-1739 by John Gregory, also titled 'John Gregory: His book'.

AUL 2206/44: 1772, *Elements of the Practice of Physic: For the Use of Students*, Balfour and Smellie, Edinburgh. AUL notation: "Interleaved copy with copious notes by John Gregory."

AUL 2206/45: 1743, 'Medical notes', including 'A proposall for a medicall society' by John Gregory.

AUL 2784/1: 1773, "A course of lectures upon the practice of medicine by John Gregory, M.D., in two volumes."

AUL 3107/1/3: 1758b, 'The state of man compared with that of the lower creation', October 11, 1758, by John Gregory, a Discourse for the Aberdeen Philosophical Society. Becomes Discourse I of 1765 edition of *Comparative View*.

AUL 3107/1/4: 1759, 'An inquiry into those faculties which distinguish man from the rest of the animal creation', August 28, 1759, by John Gregory, a Discourse for the Aberdeen Philosophical Society. Becomes Discourse II of 1765 edition of *Comparative View*.

College of Physicians, Philadelphia

CPP 10a44: 1771, 'Lectures on clinical medicine 1771-1772'. This includes material from 1773, as well.

Edinburgh University Library

EUL 2106 D: 1773, 'Lectures on the institutions of medicine, Dr Gregory Sr, 1773'.

EUL D.C.6.125: 1772, 'Lectures on the practice of medicine by John Gregory Professor of Physick in the University of Edinburgh 1772" by John Bacon. The following note appears at the beginning: "N.B. These lectures were written at Edinburgh in the years 1772 and 1773. The Manuscripts from which I copied them, were lent me by my ingenious and worthy Friend Doctor Remmet of Exeter."

EUL Dc.7.116: 1769-1770. 'The practice of physic deliver'd in the College of Edinburgh in 1769-70 by Doctor Gregory, Volume the second, wrote by Alexander Dick'.

EUL E.B. 6104 ED 1: 1744: 'Practical remarks on the sympathy of the parts of the body by the late Dr. James Crawford Professor of Medicine in the University of Edinburgh', article XLV in Medical Essays and Observations, Revised and Published by a Society in Edinburgh, Vol. V, Part II.

The Huntington Library

HL HM 17028: [dated 1765 by HL], Letter to his children from John Gregory, "Tuesday night 29th."

HL HM 17029: [dated 1766 by HL], Letter to Lord Lyttleton from John Gregory.

HL HM 17030: 1770, Letter to his daughters from John Gregory, October 6, 1770.

HL HM 17031: 1771, Letter to his daughter Dorothy from John Gregory in Dundas (?), July 24, 1771.

HL HM 17032: 1772, Letter to his daughter Dorothy from John Gregory in Edinburgh, October 22, 1772.

HL MO 949, Forbes, Dorothy (Dale), Baroness Forbes: 1760, Letter to Mr. Edward Montagu, December 20, 1760. Gregory's mother-in-law writes to request Mr. Montagu's assistance "in procuring the Professorship of Botany, in the Colledge, of Edinbr [Edinburgh] for Doctor Gregory."

HL MO 1063, Gregory, Elizabeth (Forbes): 1756, Letter to Mrs. Elizabeth Montagu, June 28, 1756. From King's College, Aberdeen [not Cambridge, as mistakenly attributed]. Gregory's wife writes to request Mrs. Montagu's assistance in gaining Gregory at appointment at St. George's Hospital in London.

HL MO 1064: n.d., Letter to Mrs. Montagu from Dr. Gregory, no date, incomplete, last portion missing. Given contents of letter, probably 1761, after death of Gregory's wife on September 29, 1761 (Lawrence 1871, Vol I., p. 157, 167).

HL MO 1065: 1766, Letter to Mrs. Montagu from John Gregory in Edinburgh, October 2, 1766.

HL MO 1066: 1766, Letter to Mrs. Montagu from John Gregory in Edinburgh, October 11, 1766.

HL MO 1067: 1766, Letter to Mrs. Montagu from John Gregory in Edinburgh, October 25, 1766.

HL MO 1068: 1766, Letter to Mrs. Montagu from John Gregory in Edinburgh, November 18, 1766.

HL MO 1069: 1766, Letter to Mrs. Montagu from John Gregory in Edinburgh, December 1, 1766.

HL MO 1070: 1766, Letter to Mrs. Montagu from John Gregory in Edinburgh, December 23, 1766.

HL MO 1071: 1767, Letter to Mrs. Montagu from John Gregory in Edinburgh, January 3, 1767.

HL MO 1072: 1767, Letter to Mrs. Montagu from John Gregory in Edinburgh, February 12, 1767.

HL MO 1073: 1767, Letter to Mrs. Montagu from John Gregory in Edinburgh, May 29th, 1767.

HL MO 1074: 1767: Letter to Mrs. Montagu from John Gregory, August 11, 1767.

HL MO 1075: 1768, Letter to Mrs. Montagu from John Gregory in Edinburgh, August 14, 1768.

HL MO 1076: 1769, Letter to Mrs. Montagu from John Gregory in Edinburgh, July 4, 1769.

HL MO 1077: 1769, Letter to Mrs. Montagu from John Gregory in Edinburgh, August 1, 1769.

HL MO 1078: 1770, Letter to Mrs. Montagu from John Gregory in Edinburgh, June 3, 1770.

HL MO 1079: 1770, Letter to Mrs. Montagu from John Gregory in Edinburgh, July 27, 1770.

HL MO 1080: 1770, Letter to Mrs. Montagu from John Gregory in Edinburgh, October 13, 1770.

HL MO 1081: 1770, Letter to Mrs. Montagu from John Gregory in Edinburgh, October 29, 1770.

HL MO 1082: 1770, Letter to Mrs. Montagu from John Gregory in Edinburgh, December 12, 1770.

HL MO 1083: n.d., Letter to Mrs. Montagu from John Gregory. Incomplete.

HL MO 1084: 1771, Letter to Mrs. Montagu from John Gregory in Edinburgh, February 24, 1771.

HL MO 1085: 1771, Letter to Mrs. Montagu from John Gregory in Edinburgh, May 3, 1771.

HL MO 1086: 1771, Letter to Mrs. Montagu from John Gregory in Edinburgh, September 21, 1771.

HL MO 1087: 1772, Letter to Mrs. Montagu from John Gregory in Edinburgh, March 7, 1772.

HL MO 1088: 1772, Letter to Mrs. Montagu from John Gregory in Edinburgh, October 22, 1772.

HL MO 1089: 1772, Letter to Mrs. Montagu from John Gregory in Edinburgh, December 14, 1772.

HL MO 1090: n.d., Letter to Mrs. Montagu from John Gregory.

HL MO 1091: 1773, Letter to Mrs. Montagu from John Gregory in Edinburgh, January 22, 1773.

HL MO 1092: 1770, "Directions for Mrs. Montagu." Diagnosis of stomach and bowel irritability and prescription for same, by William Cullen and John Gregory, September 23, 1770.

HL MO 1093: n.d. [probably 1770], medical prescription, signed "J.G.," and dated Sepr 23. This probably accompanied HL MO 1092.

National Library of Medicine, Bethesda, Maryland

NLM MS B6: 1771, 'Clinical lectures'. Ezekial Dorsey and Henry A. Ford on flyleaf.

NLM MS B7: 1768-1769, 'Lectures on the practice of physic'. Notetaker unknown.

National Library of Scotland, Edinburgh

NLS 3648: 1770, Letter to a woman not named from John Gregory in Edinburgh, January 20, 1770. The closing of this letter makes it clear that it is addressed to Mrs. Montagu: "Miss Gordon & your young friends join in their affectionate compliments to you & Mr. Montagu ...."

Royal College of Physicians of Edinburgh

RCPE Gregory, John 1:1766, 'Lectures on practice of physic', by John Gregory. RCPE Library note reads: "This may be the copy used by Gregory to lecture, 1766-73. The additions and interpolations strongly suggest original work."

RCPE Gregory, John 2: c.1766?, 'Lectures on practice of physic'. Two vols.

RCPE Gregory, John 3: 1767-1768, 'Lectures on practice of physic'. Taken by Robert Glasgow. Two vols.

RCPE Gregory, John 4: 1768-1769, 'Lectures on practice of physic'. Four vols.

RCPE Gregory, John 5: c.1770?, 'Lectures on practice of physic'. One volume. RCPE Library note: "Wanting any subsequent volumes."

RCPE Gregory, John 6: 1772, 'Lectures on practice of physic'. Taken by Joseph Camplin, two vols.

RCPE Gregory, John 7: 1771, 'Clinical lectures'.

RCPE Gregory, John 8: 1771, 'Clinical lectures'.

Royal College of Physicians and Surgeons of Glasgow

RCPSG 1/9/5: 1767, 'The Practice of Physic, As Delivered in the University of Edinburgh by Doctor Gregory, Years 1767 & 1768', by 'J. Foster, student'. One of several extant student-note versions of Gregory's Lectures (1772). Included in the present volume, pp. 71-84.

RCPSG 1/9/10: 1772, 'Notes of Lectures on the Practice of Physic begun 2d. Novr. 1772'. Included in the present volume, pp. 89-92.

Royal College of Surgeons of Edinburgh

RCSE C 12: 1771-1772: 'Cases of patients in the clinical wards. Royal Infirmary Edinburgh 1771-72', 'Dr Gregory', Two Vols.

RCSE C 36: 1771, 'Clinical lectures by Dr Gregory 1771 & Dr Cullen 1772'.

RCSE D 27: 1769, Notes taken from Dr Gregory'. One of several extant student-note versions of Gregory's Lectures (1772). Included in the present volume, pp. 85-87.

RCSE D 27: 1770, 'Lectures on the pathology by Dr Gregory Edinburgh 1770'.

RCSE E 44: n.d., no title, lectures on the practice of physic.

Wellcome Institute for the History of Medicine

WIHM 2617: 1771-1772: 'Clinical lectures' by John Gregory, notetaker unknown.

WIHM 2618: n.d., 'Lectures on the Practice of Physic', by John Gregory. One of several extant student-note versions of Gregory's Lectures (1772). Included in the present volume, pp. 67-70.

WIHM 4214: 1773, 'Clinical lectures by John Gregory'. Taken by Joshua Rigg.

B. Secondary Sources

Biographies:

Lawrence, P.D.: 1971, 'The Gregory family: A biographical and bibliographical study. To which is annexed a bibliography of the scientific and medical books in the Gregory Library,' (Ph.D. Dissertation) Aberdeen University, Aberdeen, Scotland, Two Volumes. Chapter 18 covers John Gregory, Vol. I, pp. 149-170.

McCullough, L.B.: 1998, 'John Gregory's life and times: An intellectual history', in L. B. McCullough, John Gregory and the Invention of Professional Medical Ethics and the Profession of Medicine, Kluwer Academic Publishers, Dordrecht, The Netherlands.

Smellie, W.: 1800, Literary and Characteristical Lives of John Gregory, M.D, Henry Home, Lord Kames, David Hume, Esq., and Adam Smith, L.L.D, to which are added a Dissertation on the Public Spirit and Three Essays by the Late William Smellie, Alex Smellie, et al., Edinburgh, G.G. Robinson, et al., London. "The Life of John Gregory" is found at pp. 1-118.

Stewart, A.G.: 1901, The Academic Gregories, Oliphant Anderson & Ferrier, Edinburgh. Chapter VIII is on John Gregory, pp. 100-124.

Tytler, A.F. (Lord Woodhouselee): 1788, 'An account of the life and writings of Dr John Gregory', in J. Gregory, The Works of the Late John Gregory, M.D., A. Strahan and T. Cadell, London, W. Creech, Edinburgh, Vol. I., pp. 1-85.

Truman, J.: 1995, 'The compleat physician: John Gregory MD (1724-1773)', Journal of Medical Biography, 3, 63-70.

Other Secondary Sources:

Baker, R.: 1993a, 'Deciphering Percival's Code', in R. Baker, D. Porter, and R. Porter (eds.), The Codification of Medical Morality: Historical and Philosophical Studies of the Formalization of Western Medical Morality in the Eighteenth and Nineteenth Centuries. Volume One: Medical Ethics and Etiquette in the Eighteenth Century, Kluwer Academic Publishers, Dordrecht, The Netherlands, pp. 179-211.

Baker, R.: 1993b, 'The history of medical ethics', in W.F. Bynum and R. Porter (eds.), Companion Encyclopedia of the History of Medicine, Routledge, London and New York, Vol. II, pp. 852-887.

Baker, R.: 1995, 'Introduction' in R. Baker (ed.), The Codification of Morality: Historical and Philosophical Studies of the Formalization of Western Medical Morality in the Eighteenth and Nineteenth Centuries. Volume Two: Anglo-American Medical Ethics and Medical Jurisprudence in the Nineteenth Century, Kluwer Academic Publishers, Dordrecht, The Netherlands, pp. 1-22.

Burns, C.: 1995, 'Reciprocity in the Development of Anglo-American Medical Ethics, 1765-1865', in R. Baker (ed.), The Codification of Morality: Historical and Philosophical Studies of the Formalization of Western Medical Morality in the Eighteenth and Nineteenth Centuries. Volume Two: Anglo-American Medical Ethics and Medical Jurisprudence in the Nineteenth Century, Kluwer Academic Publishers, Dordrecht, The Netherlands, pp. 135-143.

Cook, H.J.: 1995, 'Medical ethics, history of: Europe, Renaissance and Enlightenment', in W.T. Reich (ed.), Encyclopedia of Bioethics, Macmillan, New York, pp. 1537-1543.

Fissell, M.E.: 1993, 'Innocent and honorable bribes: Medical manners in eighteenth-century Britain', in R. Baker, D. Porter, and R. Porter (eds.), The Codification of Medical Morality: Historical and Philosophical Studies of the Formalization of Western Medical Morality in the Eighteenth and Nineteenth Centuries. Volume One: Medical Ethics and Etiquette in the Eighteenth Century, Kluwer Academic Publishers, Dordrecht, The Netherlands, pp. 19-45.

Geyer-Kordesch, J. 1993, 'Natural law and medical ethics in the eighteenth century', in R. Baker, D. Porter, and R. Porter (eds.), The Codification of Medical Morality: Historical and Philosophical Studies of the Formalization of Western Medical Morality in the Eighteenth and Nineteenth Centuries. Volume One: Medical Ethics and Etiquette in the Eighteenth Century, Kluwer Academic Publishers, Dordrecht, The Netherlands, pp. 123-139.

Haakonssen. L.: 1997, Medicine and Morals in the Enlightenment: John Gregory, Thomas Percival, and Benjamin Rush, Editions Rodopi, Amsterdam, The Netherlands.

Lawrence, C.: 1984, 'Medicine as culture: Edinburgh and the Scottish Enlightenment', (Ph.D. Dissertation) University of London.

McCullough, L.B.: 1978a, 'Medical ethics, history of: Britain and the United States in the eighteenth century', in W.T. Reich (ed.), Encyclopedia of Bioethics, Macmillan, New York, pp. 957-963.

McCullough, L.B.: 1978b, 'Historical perspectives on the ethical dimensions of the patient-physician relationship: The medical ethics of Dr. John Gregory', Ethics in Science and Medicine, 5, 47-53.

McCullough, L.B.: 1981: 'Justice and health care: Historical perspectives and precedents', in E.E. Shelp (ed.)., Justice and Health Care, D. Reidel Publishing Company, Dordrecht, The Netherlands, pp. 51-71.

McCullough, L.B.: 1983: 'The legacy of modern Anglo-American medical ethics: correcting some misperceptions', in E.E. Shelp (ed.)., The Clinical Encounter, D. Reidel Publishing Company, Dordrecht, The Netherlands, pp. 47-63.

McCullough, L.B.: 1984, 'Virtues, etiquette, and Anglo-American medical ethics in the eighteenth and nineteenth centuries', in E.E. Shelp (ed.)., Virtue and Medicine, D. Reidel Publishing Company, Dordrecht, The Netherlands, pp. 81-92.

McCullough, L.B., and Chervenak, F.A.: 1994, Ethics in Obstetrics and Gynecology, Oxford University Press, 1994.

McCullough, L.B.: 1996, Leibniz on Individuals and Individuation: The Persistence of Premodern Ideas in Modern Philosophy, Kluwer Academic Publishers, Dordrecht, The Netherlands.

McCullough, L.B.: 1997, 'John Gregory (1724-1773) and the invention of professional relationships in medicine', Journal of Clinical Ethics, 8, 11-21.

McCullough, L.B.: 1998, John Gregory and the Invention of Professional Medical Ethics and the Profession of Medicine, Kluwer Academic Publishers, Dordrecht, The Netherlands.

Myers, S.H.: 1990, The Bluestocking Circle: Women, Friendship, and the Life of the Mind in Eighteenth-Century England, Clarendon Press, Oxford, England.

Reich, W.T.: 1995, 'Care: Historical dimensions of an ethic of care in health care', in W.T. Reich (ed.), Encyclopedia of Bioethics, Macmillan, New York, pp. 331-336.

Reich, W.T.: 1995, 'Care: History of the notion of care', in W.T. Reich (ed.), Encyclopedia of Bioethics, Macmillan, New York, pp. 319-331.

Risse, G.B.: 1986, Hospital Life in Enlightenment Scotland: Care and teaching at the Royal

Infirmary of Edinburgh, Cambridge University Press, Cambridge.
Ulman, H.L.: 1990, The Minutes of the Aberdeen Philosophical Society, Aberdeen University Press, Aberdeen, Scotland.
Wear, A., Geyer-Kordesch, J., and French, R. (eds.): 1993, Doctors and Ethics: The Earlier Historical Setting of Professional Ethics, Rodopi, Amsterdam-Atlanta.

BIBLIOGRAPHY

Works by John Gregory:

Gregory, J.: 1765, A Comparative View of the State and Faculties of Man with those of the Animal World, J. Dodsley, London.

Gregory, J.: 1770, Observations on the Duties and Offices of a Physician, and on the Method of Prosecuting Enquiries in Philosophy, W. Strahan and T. Cadell, London. Reprinted in this volume, pp. 93-159.

Gregory, J.: 1772a, Elements of the Practice of Physic. For the Use of Students, Printed by Balfour and Smellie, for J. Balfour, Edinburgh.

Gregory, J.: 1772b, Lectures on the Duties and Qualifications of a Physician, W. Strahan and T. Cadell, London. Reprinted in this volume, pp. 161-245.

Gregory, J.: 1774, A Father's Legacy to His Daughters, W. Strahan & T. Cadell, London, J. Balfour, Edinburgh.

Gregory, J.: 1817, Lectures on the Duties and Qualifications of a Physician, M. Carey and Son, Philadelphia, Pennsylvania.

Gregory, J.: 1820, On the Duties and Qualifications of a Physician, New Edition, J. Anderson, London.

Manuscript and Unpublished Sources:

Note on Manuscript Sources. Each manuscript source is identified by a multi-letter, capitalized abbreviation, which indicates the library holding the manuscript. The code for abbreviations is the following:

AUL = Aberdeen University Library
EUL = Edinburgh University Library
HL = The Huntington Library, San Marino, California

Aberdeen University Library

Note: The information regarding titles and dates of material for the Aberdeen Philosophical Society is based on the information in Ulman, 1990.

AUL G404a: 1766, 'Address of the students of medicine to the Right Hon. Lord Provost, Magistrates, and Town-Council of the City of Edinburgh'. Printed broadside.

AUL 2206/22: 1738, 'Notebook' of 1738-1739 by John Gregory, also titled 'John Gregory: His book'.

AUL 2206/45: 1743, 'Medical notes', including 'A proposall for a medicall society' by John Gregory.

Edinburgh University Library:

EUL 2106 D: 1773, 'Lectures on the institutions of medicine, Dr Gregory Sr, 1773'.

EUL Dc.7.116: 1769-1770. 'The practice of physic deliver'd in the College of Edinburgh in 1769-70 by Doctor Gregory, Volume the second, wrote by Alexander Dick'.

EUL E.B. 6104 ED 1: 1744: 'Practical remarks on the sympathy of the parts of the body by the late Dr. James Crawford Professor of Medicine in the University of Edinburgh', article XLV in Medical Essays and Observations, Revised and Published by a Society in Edinburgh, Vol. V, Part II.

The Huntington Library

HL MO 949: 1760, Letter to Edward from Dorothy Forbes, Baroness Forbes, December 20, 1760.

HL MO 1063: 1756, Letter to Elizabeth Montagu from Elisabeth Gregory, June 28, 1756. HL mistakenly notes the original of this letter to be Kings College, Cambridge. The Gregorys were then in Aberdeen, at King's College.

HL MO 1064: n.d., Letter to Mrs. Montagu from Dr. Gregory, no date, incomplete, last portion missing. Given contents of letter, probably 1761, after death of Gregory's wife on September 29, 1761 (Lawrence 1871, Vol I., p. 157, 167).

HL MO 1066: 1766, Letter to Mrs. Montagu from John Gregory in Edinburgh, October 11, 1766.

Other Published Sources

Bacon, F.: 1875a, 'The New Organon', in J. Spedding, R.L. Ellis, and D.D. Heath (eds.), The Works of Francis Bacon, Vol. IV, Longmans, Cumpers, and Co., pp. 39-248.

Bacon, F.: 1875b, 'Of the dignity and advancement of learning', in J. Spedding, R.L. Ellis, and D.D. Heath (eds.), The Works of Francis Bacon, Vol. IV, Longmans, Cumpers, and Co., pp. 273-498.

Bacon, F.: 1875c, 'Preparative toward a natural and experimental history', in J. Spedding, R.L. Ellis, and D.D. Heath (eds.), The Works of Francis Bacon, Vol. IV, Longmans, Cumpers, and Co.

Bacon, F.: 1977, The Historie of Life and Death, with Observations Naturall and Experimentall for the Prolongation of Life, New York, Arno Press, facsimile reproduction of 1638 edition by I. Okes, London.

Baker, R.: 1993a, 'Deciphering Percival's Code', in R. Baker, D. Porter, and R. Porter (eds.), The Codification of Medical Morality: Historical and Philosophical Studies of the Formalization of Western Medical Morality in the Eighteenth and Nineteenth Centuries. Volume One: Medical Ethics and Etiquette in the Eighteenth Century, Kluwer Academic Publishers, Dordrecht, The Netherlands, pp. 179-211.

Baker, R.: 1993b, 'The history of medical ethics', in W.F. Bynum and R. Porter (eds.), Companion Encyclopedia of the History of Medicine, Routledge, London and New York, Vol. II, pp. 852-887.

Baker, R. (ed.): 1995a, The Codification of Morality: Historical and Philosophical Studies of the Formalization of Western Medical Morality in the Eighteenth and Nineteenth Centuries. Volume Two: Anglo-American Medical Ethics and Medical Jurisprudence in the Nineteenth Century, Kluwer Academic Publishers, Dordrecht, The Netherlands.

Baker, R.: 1995b, 'Introduction' in R. Baker (ed.), The Codification of Morality: Historical and

Philosophical Studies of the Formalization of Western Medical Morality in the Eighteenth and Nineteenth Centuries. Volume Two: Anglo-American Medical Ethics and Medical Jurisprudence in the Nineteenth Century, Kluwer Academic Publishers, Dordrecht, The Netherlands, pp. 1-22.

Baker, R, Porter, D., and Porter, R. (eds.): 1993, The Codification of Medical Morality: Historical and Philosophical Studies of the Formalization of Western Medical Morality in the Eighteenth and Nineteenth Centuries. Volume One: Medical Ethics and Etiquette in the Eighteenth Century, Kluwer Academic Publishers, Dordrecht, The Netherlands.

Berlant, J.: 1975, Profession and Monopoly: A Study of Medicine in the United States and Great Britain, University of California Press, Berkeley.

Burton, J.H.: 1846, Life and Correspondence of David Hume, William Tail, Edinburgh.

Bynum, W.F.: 1993, 'Nosology', in W.F. Bynum and R. Porter (eds.), Companion Encyclopedia of the History of Medicine, Routledge, London and New York, Vol. I, pp.335-356.

Climenson, E.: 1906, Elizabeth Montagu: The Queen of the Bluestockings. Her Correspondence from 1720-1761, John Murray, London, Two Vols.

Digby, K.: 1669, Of the Sympathetick Powder: A Discourse in a Solemn Assembly at Montpelier, John Williams, London.

Eccles, A.: 1982, Obstetrics and Gynaecology in Tudor and Stuart England, Kent State University Press, Kent, Ohio.

Fissell, M.E.: 1993, 'Innocent and honorable bribes: Medical manners in eighteenth-century Britain', in R. Baker, D. Porter, and R. Porter (eds.), The Codification of Medical Morality: Historical and Philosophical Studies of the Formalization of Western Medical Morality in the Eighteenth and Nineteenth Centuries. Volume One: Medical Ethics and Etiquette in the Eighteenth Century, Kluwer Academic Publishers, Dordrecht, The Netherlands, pp. 19-45.

Fletcher, J.F.: 1954, Morals and Medicine: The Moral Problems of the Patient's Right to Know the Truth, Contraception, Artificial Insemination, Euthanasia, Princeton University Press, Princeton, New Jersey.

French, R.: 1993, 'The medical ethics of Gabriele Zerbi', in A. Wear, J. Geyer-Kordesch, and R. French (eds.), Doctors and Ethics: The Earlier Historical Setting of Professional Ethics, Editions Rodopi, Amsterdam, The Netherlands, pp. 72-97.

Gevitz, N.: 1993, 'Unorthodox medical theories', in W.F. Bynum and R. Porter (eds.), Companion Encyclopedia of the History of Medicine, Routledge, London and New York, Vol. I, pp. 603-633.

Geyer-Kordesch, J.: 1993, 'Women and Medicine', in W.F. Bynum and R. Porter (eds.), Companion Encyclopedia of the History of Medicine, Routledge, London and New York, Vol. II, pp. 888-914.

Haakonssen. L.: 1997, Medicine and Morals in the Enlightenment: John Gregory, Thomas Percival, and Benjamin Rush, Editions Rodopi, Amsterdam, The Netherlands.

Hesse, M.: 1967, 'Action at a distance and field theory', in P. Edwards (ed.), Encyclopedia of Philosophy, Macmillan, New York, pp. 9-15.

Hoffmann, F.: 1749, Medicus Politicus, sive Regulae Prudentiae secundum quas Medicus Juvenis Studia sua et Vitae Rationem Dirigere Debet, in F. Hoffmanni, Operum Omnium Physico-Medicorum Supplementum in Duas Partes Distributum, apud Fratres de Tournes, Genevae.

Hoffmann, F.: 1971, L.S. King (trans.), Fundamenta Medicinae, MacDonald, London.

Hume, D.: 1978, P.H. Nidditch (ed.), A Treatise of Human Nature, 2nd. ed., The Clarendon Press, Oxford, England. Based on 1739-1740 edition.

Johnston, D.B.: 1987, 'All honourable men? The award of irregular degrees in King's College and Marischal College in the eighteenth century', in J.J. Carter and H.J. Pittock (eds.), Aberdeen and the Enlightenment: Proceedings of a Conference Held at the University of Aberdeen, Aberdeen University Press, Aberdeen Scotland, pp. 136-145.

King, L.S.: 1970, The Road to Medical Enlightenment, 1650-1695, MacDonald, London, American Elsevier, Inc., New York.

King, L.S., 1978, The Philosophy of Medicine: The Early Eighteenth Century, Harvard University Press, Cambridge, Massachusetts.

Lawrence, C.: 1984, 'Medicine as culture: Edinburgh and the Scottish Enlightenment', (Ph.D. Dissertation) University of London.

Lawrence, P.D.: 1971, 'The Gregory family: A biographical and bibliographical study. To which is annexed a bibliography of the scientific and medical books in the Gregory Library,' (Ph.D. Dissertation) Aberdeen University, Aberdeen, Scotland, Two Volumes.

Leake, C.: 1927, Percival's Medical Ethics, Williams and Wilkins, Baltimore.

Loudon, I.S.L.: 1993, 'Childbirth', in W.F. Bynum and R. Porter (eds.), Companion Encyclopedia of the History of Medicine, Routledge, London and New York, Vol. II, pp. 1050-1071.

McCullough, L.B.: 1978, 'Historical perspectives on the ethical dimensions of the patient-physician relationship: The medical ethics of Dr. John Gregory', Ethics in Science and Medicine, 5, 47-53.

McCullough, L.B.: 1983: 'The legacy of modern Anglo-American medical ethics: correcting some misperceptions', in E.E. Shelp (ed.)., The Clinical Encounter, D. Reidel Publishing Company, Dordrecht, The Netherlands, pp. 47-63.

McCullough, L.B.: 1984, 'Virtues, etiquette, and Anglo-American medical ethics in the eighteenth and nineteenth centuries', in E.E. Shelp (ed.)., Virtue and Medicine, D. Reidel Publishing Company, Dordrecht, The Netherlands, pp. 81-92.

McCullough, L.B.: 1998, John Gregory and the Invention of Professional Medical Ethics and the Profession of Medicine, Kluwer Academic Publishers, Dordrecht, The Netherlands.

Myers, S.H.: 1990, The Bluestocking Circle: Women, Friendship, and the Life of the Mind in Eighteenth-Century England, Clarendon Press, Oxford, England.

Nutton, V.: 1993, 'Beyond the Hippocratic Oath', in A. Wear, J. Geyer-Kordesch, and R. French (eds.), Doctors and Ethics: The Earlier Historical Setting of Professional Ethics, Editions Rodopi, Amsterdam, The Netherlands, pp. 10-37.

Porter, D. and Porter, R.: 1989: Patient's Progress: Doctors and Doctoring in Eighteenth-Century England, Stanford University Press, Stanford, California.

Porter, R.: 1987, 'A touch of danger: The man-midwife as sexual predator', in G.S. Rousseau and R. Porter (eds.), Sexual Underworlds of the Enlightenment, Manchester University Press, Manchester, pp. 206-232.

President's Commission for the Study of Ethical Problems in Medicine and Biomedical and Behavioral Research: 1981, Defining Death: Medical, Legal, and Ethical Issues in the Determination of Death, U.S. Government Printing Office, Washington, D.C.

Ramsey, P.: 1970, The Patient as Person: Explorations in Medical Ethics, Yale University Press, New Haven, Connecticut.

Rather, L.J.: 1965, Mind and Body in Eighteenth Century Medicine: A Study Based on Jerome Gaub's De Regimine Mentis, University of California Press, Berkeley.

Reid, T.: 1990, K. Haakonssen (ed.), Practical Ethics: Being Lectures and Papers on Natural Religion, Self-Government, Natural Jurisprudence, and the Law of Nations, Princeton University Press, Princeton, New Jersey. Based on manuscript sources.

Risse, G.B.: 1986, Hospital Life in Enlightenment Scotland: Care and teaching at the Royal Infirmary of Edinburgh, Cambridge University Press, Cambridge.

Royal College of Physicians of Edinburgh: 1882, Historical Sketch and Laws of the Royal College of Physicians of Edinburgh, from its Institution to August 1882, Royal College of Physicians, Edinburgh.

Shakespeare, W.: 1952, 'Macbeth', in S. Barnet (ed. and intro.), William Shakespeare, Four Great Tragedies: Hamlet, Othello, King Lear, Macbeth, each play paginated separately.

Smellie, W.: 1800, Literary and Characteristical Lives of John Gregory, M.D, Henry Home, Lord Kames, David Hume, Esq., and Adam Smith, L.L.D, to which are added a Dissertation on the Public Spirit and Three Essays by the Late William Smellie, Alex Smellie, et al., Edinburgh, G.G. Robinson, et al, London.

Sperry, W.: 1950, The Ethical Basis of Medical Practice, P.B. Hoeber, New York.

Steuart, R.: 1725, The Physiological Library. Begun by Mr. Steuart, and Some of the Students of Natural Philosophy in the University of Edinburgh, April 2, 1724. And Augmented by Some Gentlemen; and the Students of Natural Philosophy, December, 1724.

Stewart, A.G.: 1901, The Academic Gregories, Oliphant Anderson & Ferrier, Edinburgh.

Tong, R.: 1993, Feminine and Feminist Ethics, Wadsworth Publishing Company, Belmont, California.

Truman, J.: 1995, 'The compleat physician: John Gregory MD (1724-1773)', Journal of Medical Biography, 3, 63-70.

Tytler, A.F. (Lord Woodhouselee): 1788, 'An account of the life and writings of Dr John Gregory', in J. Gregory, The Works of the Late John Gregory, M.D., A. Strahan and T. Cadell, London, W. Creech, Edinburgh, Vol. I., pp. 1-85.

Ulman, H.L.: 1990, The Minutes of the Aberdeen Philosophical Society, Aberdeen University Press, Aberdeen, Scotland.

Underwood, E.A.: 1977: Boerhaave's Men at Leyden and After, Edinburgh University Press, Edinburgh, Scotland.

Waddington, I.: 1984, The Medical Profession in the Industrial Revolution, Gill & Macmillan, Dublin.

Wear, A.: 1987, 'Interfaces: Perceptions of Health and Illness in Early Modern England', in R. Porter and A. Wear (eds.), Problems and Methods in the History of Medicine, Croom Helm, London, pp. 230-255.

Wear, A.: 1993, 'Medical ethics in early modern England', in A. Wear, J. Geyer-Kordesch, and R. French (eds.), Doctors and Ethics: The Earlier Historical Setting of Professional Ethics, Rodopi, Amsterdam and Atlanta, pp. 98-130.

Wear A., Geyer-Kordesch, J., and French, R. (eds.): 1993, Doctors and Ethics: The Earlier Historical Setting of Professional Ethics, Rodopi, Amsterdam-Atlanta.

Whytt, R.: 1765, Observations on the Nature, Causes and Cure of those Disorders which have been commonly called nervous, Hypochondriac or Hysteric .... 2nd. ed., corrected, Balfour, Edinburgh.

Wilson, C.: 1995, The Invisible World: Early Modern Philosophy and the Invention of the Microscope, Princeton University Press, Princeton, New Jersey.

JOHN GREGORY

WHETHER THE ART OF MEDICINE AS IT HAS BEEN USUALLY PRACTICED HAS CONTRIBUTED TO THE ADVANCEMENT OF MANKIND

Citation:
Gregory, John
Aberdeen Philosophical Society, Question 59, July 12, 1762 (Ulman, 1990, p. 193): "Whether the Art of Medicine as it has been usually practiced has contributed to the advancement of mankind."
Aberdeen University Library: MS 37, 1762, pp. 162r-163v, continued at p. 13r.

Comment:
This document uses many contractions. To reproduce them here with transcriptions in brackets, as I have done on some of the other documents with less use of contractions, would unnecessarily burden the reader. For example, in the statement of the question, only one noun appears uncontracted, 'Art', along with 'of', 'as', 'it', and 'to'. I retain the ampersand where it appears in the original. The pages are numbered on recto side only. Page numbers appear in brackets at end of original page. Original spellings are retained.

Acknowledgment: This document appears with the permission of the University Library, University of Aberdeen, Colin McLaren, Head of Special Collections and University Archivist.

Text:

Q: Whether the Art of Medicine as it has been usually practiced has contributed to the advancement of mankind.
We cannot suppose any state of mankind where the art of medicine would be unnecessary. In a state of nature we have no reason to doubt, but there would be as much exemption from disease as all other animals are in the

natural state – But even here they must be subjected to Disorder from extensive injury & when one found himself in distress, he would be prompted by reason as well as instinct to seek relief. – There is this remarkable diference between an animal and an inanimate machine – When any disease happens in the former, there is a certain internal process carried on by nature, the principles of which we are in a great measure strangers to, by which they endeavor to remove the disorder or to suppress any want occasioned by it – We have instances of this in the cases of fractured bones, the incarnation of wounds, and the inlargement of one kidney when the other is destroyed – In a common machine there is no internal power analogous to this – Bŭt the efforts of nature to cure a disease or relieve a patient may in some cases without the assistance of Art, be inefficient and in some cases pernicious, e.g., dislocation fractures before they are sett & children presenting wrong [for delivery] – Sometimes we cannot safely allow nature her course in carrying off a disease as in the case of an internal infection which nature would generally carry off by a supposition satisfactory to the patient – The efforts of nature are sometimes so violent as a require a check & sometimes so feeble as to stand in need of cordials & an additional stimulus – In some cases nature makes no sensible effort for her relief – These facts which cannot be disguised lay the foundation for the art of medicine – Men no doubt have suffered greatly from the unskilled practice of the art – But whether it has been better for mankind that every disorder has been left to nature, without any assistance of medicine is a question not easy to ascertain, as we have not one instance of a nation where the art was not practiced – No people would be so stupid or obstinate as to leave a dislocated arm to nature – The only question then can be to settle the proper limits between the art and the operations of nature in the cure of disease, a question which requires the deepest knowledge of both resolve – Tis a melancholy reflexion however to consider that there is too much reason to suspect that in some whole classes of diseases, many more have perished by medicine than would have done if they had been left intirely to nature and the direction of instinct – It is too probable that the most of cases of some species of fever and particularly of the small pox before the time of Dr Sydenham[1] – Instinct in those fevered directed the sick to require cool and fresh air, cold water small beer or acidulated drink – Physicians again shut them up in close hot rooms etc. so that the poor patient was boiled to death, or if he recovered, might truly be said to have escaped with a δία πορος – But instead of entering into that useless question whether [medicine] has contributed most to the benefit or hurt of mankind, let us enquire whether it is most beneficial to have it confined to & practiced as it usually is by a sett of men who live by it as a profession or to leave it large to be studied, like all other branches of natural philosophy by those who love science, but have a fortune or some other profession to support them – It is said that medicine is so complicated a science that it requires the whole of [162r] of a persons time & application to understand it, & yet little progress has been made notwithstanding the labours of so many ingenious and learned men entirely devoted to its cultivation is adduced as a proof of its difficulty & intricacy – It is said therefore that if there were no such distinct profession as medicine, the art would rather decline than be improved, & and the consequences of people studying and practicing Physic, who have not been regularly bred to it, would be multiple quacks and lessening the confidence and implicite faith in the physician which is convenient for the patients own sake as well as the doctors – This reason has appeared so powerful to the faculty that we find no profession which has watched with so jealous an eye over intruders or which has treated them with so much abuse & ridicule even when it was apparent the intrusion was only owed to motives of humanity – It would not be candid to ascribe this to any sordid views – Enlarged knowledge produces a liberal & unsuspicious spirit & there is no profession that can boast of more men of learning, ingenuity, & genteel education than physic – There is however some reason to suspect that the way in which Physic has been practiced has not been so beneficial to mankind nor contributed so much to the advance of the art, as if it had been left more open & not confined to a particular set of men who lived by it as a profession – In the first place universities where medicine is usually taught seem not to be well calculated for the advancement of any kind of science – For the conveiency of teaching every science is usually taught in the synthetic method where general principles are laid down & particular facts are mentioned only in so far as they can be deduced as corollaries from the principles – But the natural and only genuine method of acquiring sciences is in the analytical way, which is the very reverse of the former, where the mind proceeds from particular facts to established general principles – Though on a superficial view it does not seem of much importance in which way science is acquired, it will appear on a nearer view to be of great consequence in influencing the mind and direction of ones further character and enquiries – Medicine as taught in colleges is digested into a regular and compleat system – In this view it is beheld by the young student, who embraces the theories with the same facility & unsuspecting confidence that he would do facts – He understands the causes of all diseases & the operations of all medicines – His mind is at ease in having allways sure and fixed principles to rest upon – In the mean time the art can receive no improvement from him, as he does not imagine it stands in need of any – If a patient dies he is quite satisfied everything was done for him, that art could do – It is difficult for men to give up favourite opinions, the children of our youth, to sink from a state of security & confidence, into one of suspence and scepticism – Accordingly, few physicians change their principles and very seldom their practice – These vary like other fashions and like them are universally adopted, while they do prevail. Boerhaave wrote his Aphorisms when a very young man – He published many editions of them, & the last when he was very old, differs in nothing material from the first – The same [162v] may be said of Dr Mead[2], whose Medical precepts, wrote down when he was near, 80, are remarkable for nothing but the purity and elegance of the title – He would have wrote precisely such a book at 25 – It was a severe thing which Dr Garth[3] said of his brethren when he was dying – One of his patients asked him, what physician he would advise him to employ – The Dr very gravely bid him always send for the nearest – Physicians are generally accused of want of faith in religious matters – Surely their education in their own profession does not dispose them to infidelity – A physician who does not see the weakness & futility of the best of a physical system is certainly well prepared to embrace any system of religious faith that can be offered to him – If we can enquire now into the effects produced on the mind by acquiring knowledge in the slow method of observation & induction from experiments, we will find him very different – The mind here gains a close & accurate attention to facts (having nothing else to trust to), slow in forming principles from these facts & diffident of them when formed, instead of him being assuming & dogmatical, is modest & scpetical – A Physician of this stamp never loses a patient but he is secretly laments his ignorance of the proper means of having saved him, which he blames rather than the disease being incurable in itself – This naturally stimulates to the improvement of knowledge, both from a love to science & from humility, and a principle of conscience – We own that the phylosophical mind, if not carefully attended to, may become very deterimental in the practice of physic & indeed in all the practical parts of life, by making a physician timid & irresolute when he must act – But the true phylosophy leads [the physician] to be diffident & cautious in forming opinions, but when there is occasion to act, to be quick in forming a resolution, & resolute & fearless in putting it into execution – But, secondly, it would seem that the confinement of the practice of physic to a sett of men who have no other method of subsistence, is unfavourable to the practice of the art, beacuse the mind of improvement is often checked by and is sometimes incompatible with the necessary attention to his private interests – Physicians are neither better nor worse than the rest of mankind – They are influenced by the same general motives of action – A physician, when he sets out in life, quickly perceives that the knowledge most necessary to procure him a subsistence, is not the knowledge of his profession, what he finds most essential to that purpose are the arts of deceiving mankind, into an opinion of his understanding by an appearance of solemnity, & importance in his whole deportment, and the various arts of flattery & dissimilitude, views very different from the pursuit of genius & science – He can expect no patrons to his real merit, because none are judges of it, but a few of his profession, whose interest it is to have it concealed – If he attempts to show the weakness of the fashionable system or to introduce an alternative in the practice, the whole faculty are alarmed, their vanity is piqued in having their opinions, which they thought perfectly well established, brought into question & exposed by a young man & their interest is evidently concerned to crush him as fast as possible – In the mean time the effect of every deviation he made from the common practice is anxiously watched, all his prescriptions must remain in the apothecarys file to stand in judgment against him & upon any miscarriage the outcry is raised & propagated with the utmost malignity – Now as physic is as yet only a conjectural art, no improvement can ever be attempted in it, without a risk of bad success, & the opportunity of misapprehension not easily to be refuted, are so many that few it can well be imagined will venture the consequence – Indeed in the pursuit of interest or ambition a man does not hate another because he is engaged in the same race [163r] but let him only keep his proper distance behind & he will wish him very well – Sometimes it happens that a man of mind & resolution rises above all these difficulties, but it happens unfortunately that the word of assurance does not so often accompany literary merit, as diffidence & modesty – What is said above is meant only to describe the characteristic mind of a class of men, but not at all to include all the individuals of the profession – There is a consequent dignity that generaly acompanies genius wherever found, which renders the possessor equally impervious to the feelings of envy, & all the low arts of dissimulation – But not to insist on further on arguments to prove that no considerable improvement in the art of medicine can be expected from physicians on their present footing, we will only observe the fact that it appears from the history of medicine that the improvement in it has been extraordinarily seldom owed to physicians, nay it is most extraordinarily that they have been generally opposed by them, with great keenness & acrimony & never adopted by them without a great struggle – We could give instances in almost every improvement of modern medical practice: practices in the cure of blisters, opates, cortex Pervuianus [Peruvian bark], antimony, mercury & all powerful chemical remedies – The inventors of these from the days of Paracelsus down to Dr Ward[4], have been held by the faculty in contempt and detestation – The disfavour of all such men as were not regular physicians were never examined with the candor and impartiality which their importance and success required – That from some men the greatest discoveries are to be expected – They are not fettered by regular physicians, as they seldom can suffer much either in their interest, or reputation by the bad success of their experiments – It was from the lowest artificers and strolling chemists & not from the schools of philosophers that Mr Boyle drew that large & useful collection of factors with which he has enriched Physic – Even Dr Sydenham[1], the only physician before our time to whom the practice of physic owed an improvement, was looked upon by most of his brethren in his own time as little better than an ignorant quack, & it was one of Boerhaave's greatest merits in physic that he was the first who made that great man's character known and respected among physicians – These observations make it probable that little improvement can be expected in medicine while it is monopolized by a sett of men who have a separate and distinct interest from that of the art, & that if it were cultivated by men of sense & knowledge who have no other view but the love of the science & the benefits of mankind, it would make advances correspondent to other branches of natural knowledge – A strange fate seems in many instances to influence the condition of mankind – They are tenacious and jealous of their liberty & property – They still make a trouble about priestcraft, & the authority claimed by priests over their consciences while the priests live quietly & inoffensively & [not] troubling themselves about any bodys conscience, & yet they will readily trust their life & health with the most implicit confidence into the hands of any fool of a physician, whom accident throws in their way – Such a knowledge of medicine as would enable one to know every thing related to the preservation of health & the cure of disease which physicians commonly knew can be attained by a man of sense & letters, with no great study & application – Custom indeed here makes it necessary for every physician to go through allmost every branch of natural philosophy & history before he is supposed qualified, to practice physic – It is likewise expected he should be master of greek & latin & most fashionable modern languages – The present fashion they think does not require a knowledge of Arabic, the Arabic physicians being in disgrace – In short Haller writes 2 large 4to [quarto] volumes on the method of study of physic, but in truth this is all pomp and affectation – We readily own that all parts of natural knowledge are connected & serve to illustrate one another – But it does not follow from this that a person in order to understand one science must study all the [163v; text ends] [the text continues at 13r:] the rest. Human life is so short for that – a man grows old before he arrives at the great object of his pursuit – the art of curing diseases requires the whole and immediate attention of a physician, to that single object – We cannot afford time for the study of any branch of science but what directly & immediately throws light on the Art – All solitary and uncounted facts, that point at no principles, and lead us to no conclusions, are as useless encumbrances on the memory – We shall exemplify this by an instance from Chemistry – A student examines with great care and attention all the varieties of the several metalic ones, that are displayed by the Professor, & this is called the study of Physic – [contracted writing ends here] Now this is certainly a very necessary study to one who proposes to work in Mines, but its connexion with Medicine is wonderfully remote – Tis a physicians business to be acquainted with the chemical history of the animal humours of all the Medicines he makes use of, & of the principles of Pharmacy which is intirely a branch of Chemistry – But this branch of chemistry the only important one to a Physician – has only been attended to with in these very few years – The best chemical history of the animal humours is to be found in some late thesis of students of physic at Edinburgh – The science of Pharmacy till the late Reformation in our Dispensatories was in so rude & wretched a state – founded on no principles & often inconsistent with all Principles both of Chemistry & Comon Sense, as was a disgrace to any sett of men, who called themselves a learned profession – By the late Reformation which was the work of a few men, this Art is now reduced to some elegance & simplicity – But it is as fact, that the Principles of Pharmacy on which the Reformation was made are still known but to a very few physicians, & that the

Reformation is far from being generally adopted – Many observations of [text ends][13r]

JOHN GREGORY

LECTURES ON THE PRACTICE OF PHYSICK, 1767-1768

Citation:
Gregory, John
Lectures on the Practice of Physick, 1767-1768, taken by Sir Charles Blagden. Wellcome Institute for the History of Medicine: MS 2618, 1767.

Comment:
Volume I of Holograph notes taken by Sir Charles Blagden (1748-1820), when a student at Edinburgh University. The notes comprise ten volumes and include lectures by William Cullen Vol. I is Wellcome Ms. 2618. Written on recto pages only, with no page numbering. Page numbers are assigned and appear in brackets at the end of each original page. The text is more in the style of notes, rather than verbatim. Contractions are transcribed on their first appearance. Original spellings are retained.

Acknowledgment: I gratefully acknowledge the assistance of the Wellcome Institute for the History of Medicine Library and Richard Aspin, Curator of Western Manuscripts at the Library, in providing me access to this manuscript material.

Text:

Lecture 2, Nov. 2
Observations on the office & duties of a physician, genius & moral qualities.
Humility, patience, secrecy & honour.
Decorums: what tends to support his dignity. His behaviour. Mention the course of education necessary for proper dignity.
1. Genius, understanding, temper.
None requires so comprehensive a mind as medicine. In others certain standards & practices only. Assiduous application is necessary and mem-

ory: little exercise of ingenuity. All reasonings must be referred here to temporary determination.
In medicine no standard authority in doubtful cases, rest on his own judgment, appeal to experience. We must make a separation between the just & improper he learns. This done difficult to apply to practice. For diseases are complicated tho he learns them separately. Hence accuracy of judgment. He thinks he knows all diseases at first & can cure. But this is false pride. [1] A physician must combat with prejudices of the world, his brethren etc.; hence much common sense & knowledge of the world. Sudden emergencies occur which may embarass, such occasions call for the quickest discernment & resolution. By perverseness his temper may be ruffled. Hence be resolute and have a command of temper.

MORAL QUALITIES

Humanity, sensibility of heart exciting to relief; produces attention to the most minute circumstances. Procure confidence of the patient. Men of compassionate tempers acquire composure can feel what is amiable in pity without being enervated. Often affected but easily discovered: by different behaviour in high & low life. Worthy men show it less to those in high life. A gentle & humane temper attends vigor of mind.

If he is rigid his directions will not be followed & deviations concealed. Hence not know the true state. The government should be absolute but he should prescribe such laws that will be obeyed. It is necessary to suppose a proper dignity. [2]

Patients suffer nervous complaints. give great trouble; This is a real disease of the constitution. Disorders in the imagination may be the object of a physic$^{n's}$ [physician's] attention. If he treats slightly or with unfeasible ridicule he shocks; if too seriously he rivets. Strike the medium: introduce amusing or interesting discourse, or good natured ridicule. Some treat sarcastically to the poor; foster them in rich. very diff^{t} [different] behavior at beginning or advanced practice: A physician is employed because it is fashionable to do so. He can know the private transactions of family. see pevishness etc marks of broken temper – Most profound secrecy where the women are concerned. Certain maladies, the concealment of which is of consequence. Has many opportunities of seducing; if he does mean, unworthy. Some eminent physicians prescribe as well drunk as sober. Showed they practised by rote. Chearful glass often enlivens invigorates – Candour openeth conviction. Obstinacy often owing to self conceit [3] Knowledge inconsistent with all conceit – Often too proud to acknowledge in error. Must not discover to Patient lest lose confidence.

Decorum decency & propriety are undetermined words. partly founded in nature & partly in caprice etc. The first immutable, the latter necessary.

Allow every indulgence consistent with safety. Often propose a remedy, may suggest what wo^{d} [would] not have occurred. hence use it; some will pos: [possess] an insolent tyranny. Proposal to be made with deference, it approve declare, if not, tell so as to show by judgment. He must not complain of his advice not being followed.

Necessary not to tell truth with regard to life sometimes. When no provision, warn the patient of his danger. Never conceal real situation from relations. This is a disagreeable duty, but indiffensible. If he recovers, joyful, if not, less shock. Not leave when life despaired. Come as a friend. Physician & clergyman mutually assist each other. [4] Enthusiast may injure a patient.

Must be different fees from different states. When he does not take fees must be more attentive. A physician never involve his patient in private piques. If his mind is not open to conviction he sho^{d} decline consultation. A medicine may occur to one w$^{c.}$ [which] would not to another, & advice give confidence. If no honour, secrecy to be observed consultations rather detriment. Appeals to the public are detrimental to all parties & to the faculty in general. Do every thing that is not morally criminal for his patient.

Great disputes have arose about the boundaries of physick & surgery. Sometimes ancient phys^{ns} performed manual oper^{ns} [operations] themselves. Sometimes by slaves. Among many of the moderns united, but often separated. Class surgeons with Barbers, not reckoned liberal. All diseases are so connected that cannot understand one without the other. If gangrene or broken leg physic must judge if operation necessary, or if medicine be tried. [5]

Operating if practised alone, wo^{d} be more perfect. Surgeons act as physicians in ordinary, or Apothecaries. Hence none to the greatest advantage. Physicians must often depend on Apothecary, ignorant illiterate. Every branch respectable. Title of Dr or not of no consequence. Apothecary if he has the knowledge, he is a phys. [physician]. Man of candour never take advantage of titles. He will despise distinction founded in vanity, caprice, etc. – Formality has been much regarded.

Sometimes in nature, sometimes in caprice. In judges & magistrates pomp of dress necessary. But in medicine no natural propriety. no respect or authority to his office. External form are insidious snares to mankind. Supplant more worthy expose to ridicule or contempt. Hence no obligation but fashion, which common sense & prudence render necessary to regard: But if he may dress as other people without contempt, avail himself of it. [6]

Must not indulge in unnecessary over refined delicacy: always sho[d] give place to duty or humanity: Nothing neglected which may contribute to the ease of his patients.

Servility of manners, to people or rank or fortune; a most humbling circumstance. Pay the respect w[ch] order of society requires. Superior parts seem a nat[l] [natural] foundation of pride elation; this is a pardonable weakness if not accompanied with insolence. Men of retirement & study aukward. Liberal manners & liberal spirit not confined to any.

Nostrums, as having air of mystery catch the minds of the people. Awe & marvelous operate on the imagination. A divulged nostrum is forgot. By this means only, 'tis said, a remedy can be recommended or fairly tried. Nostrums do much more hurt than good. Stop science, impose on the public, in the hands of illiterate men. Where men of knowledge have them, not the same harm. [7]

The most eminent of our faculty have had a regard to religion; but not bigots; impatient of controul on consciences this moderation called impiety. Exasperated have expressed themselves unguardedly. Natural religion rejected unhinges the bonds of society. Call Hypocrites or fools which imputations youth dread. The most unworthy men have been best defenders of religion. Most credulous often dispute religion. Physicians see the most gay the happy plunged in distress; hence reverence that religion which can support. Physicians of politeness & good manners sho[d] prevent proposing their doubts; but more especially those of kindness compassion; not to take away from the patient bidding adieu to sublimary pleasures his only last support. [8]

JOHN GREGORY

LECTURES ON THE PRACTICE OF PHYSICK, 1767-1768

Citation:
Lectures on the Practice of Physick, 1767-1768
reported by J. Foster
Royal College of Physicians and Surgeons, Glasgow: MS 1/9/5, 1767.

Comment:
The text contains very few contractions, which are included as in original with transcription at first occurrence. The ampersand is included where it occurs in the text. Page numbers appear in brackets at the end of each page of the original text. Where it appears (e.g., at end page I), the first word of the next page is included; hence, there is some repetition. The text is numbered in Roman capitals through p. XXVII then continues with p. I. Emphasis throughout original. Original spellings are retained. Original underlinings are also retained. They appear either for emphasis or to mark, at the bottom of a page, the first word of the following page.

Acknowledgment: This document appears with the permission of the Library and Archive Committee of the Royal College of Physicians and Surgeons of Glasgow, James Beaton, College Archivist and Librarian.

Text:

Preliminary Lectures
Gentlemen

My design here is to explain the Practice of Physic, which is the art of preserving health, prolonging Life & Curing Diseases.

For Which Branch I hope that all your former Colledges & other Branches of Medicine have sufficiently prepared you. –

I shall first then as is the Custon give you some Preliminary Lectures. –

The Utility of this art has never been called in question or doubted of, For what can be more usefull than that art which teaches to preserve health, by Preventing and Curing Diseases. –

Much Wit indeed, & criticism has at all times & in all ages, been exerted against our profession, but that has always been chiefly aimed at certain ignorant Physicians Professors in the art, more than at y^{e}. [the] Science of Physick itself. Some of its professors are Gentlemen of Honour, Modesty, candour, integrity & cet. [et cetera] but in this as in other Societys of men, these do not make the most numerous part, of it. Others again have professed this art to cloak ignorance, & cover their [I] their poverty. – Physicians are much Raged against, thus in a dramatic performance, they are always treated as solemn coxcombs & fools, but this is levell'd at Particular persons Professors of the art, & not at Physicians in general. –

If a Gentelman is for studying a liberal art he cannot choose a better than y^{t}. [that] of Physick, which is one of y^{e}. most liberal arts y^{t}. can be professed.

In order to understand aright y^{e}. Practice of Physick, a knowledge of y^{e}. natural arts is necessary, such as Mathematics, Mechanics, natural Phylosophy, Metaphysicks & also the other Branches of Medicine, as Anatomy, Chymistry, Materia Medica, & there is an absolute necessity for an aquaintance w^{t}. [with] y^{e}. Latine, Greek, & French Languages.

No Art requires a more liberal acomplishment than that of Physick. –

This Art has flourished most in y^{e}. British Dominions, where it has been more regarded than in any other places of the Universe. –

Besides the utility of this art to mankind, it 1st. [first] presents us w^{t}. an ample feild for improvement and 2dly. [secondly] furnishes us w^{t}. an extensive feild for [II] exerting our humanity, Charity & cet. it is an excellent Subject for y^{e}. Human mind to work upon. To do mearly as one man should do to another, also to display his good nature, patience, honesty, & Charity towards his Brethren of Mankind. –

Physicians have often been considered as hard hearted, for this reason, Because they have so often & so many oppertunities of Considering & being involved in human misery, that at last they grow insensible to any Human affection, but this Supposition is often false & ill grounded, yet on the other hand a Physician ought not to be too tender hearted, as it will be both much to his own prejudice, & his patients hurt.

Medicine has by some been divided into two parts. –

1st. An Art for preserving y^{e}. Health of Mankind,

2dly. For Supporting its Professors. but this is an unjust argument, as many Gentlemen mearly for y^{e}. sake of y^{e}. utility of the art, have professed it, & not from any views of interest [? part of page missing] at all. – – In [III] In the prosecution of this Subject then, I shall first consider the Qualifications necessary to fit a man to be a Physician. –

First then of the Genius, Nature, & other Qualifications of a Physician. –

There is not as in other arts any established Authority in Medicine, every man must depend on his own judgement in affairs.

Physicians find it necessary to distinguish between important facts, & bare Theoretical Reasoning, in order to attain to a true knowledge in Medicine it is necessary that every Disease should be considered separately & by its self. –

A student w^{n}. [when] first setting out in y^{e}. world in the pride of his heart fancies every disease must yield before him, it will be important however for his patients as well as himself, if a little more Experience humbleth this pride, & shews him that there are many cases, and Diseases which he knows nothing about. –

A Physician should have acute solid judgement, a Clear happy Genius as he has not only his own faults to conquer but [? part of page missing] must [IIII] also struggle w^{t}. the prejudices of his patients. and of his Brethren in y^{e}. Faculty & of others; and hence it is necessary for a Physician to have a good knowledge of y^{e}. world, sometimes unforeseen events may fall out in a Disease which may very much incumber, & embarass a young hot Spirited Student, but he must be aware of this, & let it not discompose him, therefore a Physician should be possest of great presence['of' crossed out] Steadiness, & Composure of mind.–

Next y^{e}. Moral Qualifications necessary for a Physician. –

First Sensibility of heart, as a natural feeling for our fellow Creatures. – We should have a Sympathy for them, Sympathy will often produce things to our advantage which money never could do, it will produce y^{e}. attention & good will of y^{e}. Patient to y^{e}. Physician which he should always Cultivate, whereas a Brutal harsh Physician, by the rudness of his manner makes the patients heart sink within him, however this Sympathy among Physicians is said [V] is said to be often affected, but if such is ever the Case, the affectation is very easily seen thro', thus some Physicians affect to Sympathise very much w^{t}. great folks, people of high rank, & even in a manner to nurse them, whereas they intirely neglect those of Low rank, But the real will never distinguish between those of High &

Low rank, or if it does it will even show more assiduity in Low rank, this Sympathy in some Measure consists in a certain Gentleness and flexibility of y^{e}. Physician to his patients, not at all strick [strict] or severe as to diet or otherwise, & in this manner he will find his orders much more regularly, & punctually obeyed, if he is extremely rigid, & severe in his regimen, they then will be much more apt to disobey him, & deviate from his orders, & then perhaps he will be accounting for the surprising effects of Medicines w^{h}. [which] were every day thrown over the window.

Now patients often put the patience of a Physician to a very severe trial, therefore he needs a steady mind.

It is the Physicians duty to do every thing in his Power for y^{e}. Patients relief, & if there is any dist [fragment of page missing] nce [VI] in the patients mind, it is the duty of the Physician to cure Distempers of y^{e}. mind, as well as of the Body so fas as he is able. –

The Physician should not have too minute a concern nor be too anxiously sollicitous about every triffling circumstance w^{h}. may happen as this rather increases the Disease by attracting too much the attention of y^{e}. patient, nor on y^{e}. other hand should he pass over things too slightly, as of no Moment, w^{h}. perhaps might have been of great use if attended to, but he should endeavour to strike a due Medium betwixt these two. –

It is said among the Tribe, that there is no disease so Lucrative as those of the Nervous kind, as most people in High rank are troubled generally w^{t}. those complaints. –

The Young Physician w^{n}. beginning y^{e}. world is affable, Polite, Humane, & good natured to his patients, but in a short time, w^{n}. he becomes a little more acquainted w^{t}. y^{e}. world he grows naughty [haughty?], Proud spirited, Daring, & so as to boast even that no person dare die without his permission. –

No person more than the Physician is allowed or has oppertunities of observing people in private Life, What a great charge in this committed to the Physicians [VII] Physicians honour, the observation of private families in their different Situations, which ought to be kept with the most inviolable Secrecy, as the Peace & happiness of many families depends entirely upon it, & upon the Discretion of the Physician, There are also many Circumstances w^{h}. women especially, from y^{e}. Extrem Delicasy of their Nature, w^{d}. [would] have conciled [concealed], & who depend intirely upon the Physician for this.

Again a Physician who is a man of Galentry injoys the best opp^{tys}. [opportunities] of any person what ever for Seducing his patients, but if he does such a thing he should be looked on a base vile wretch, & not fitt to be trusted.

It is said of some Physicians that they prescribe as well or better w^{n}. Drunk than Sober, if Such were the Case, it shows these persons of no worth or Learning, & that they generally prescribe by rot [rote].

Another Quality necessary for a Physician is Candour, this lays him open to Correction, when in an error, the ill treatment of a disease may be owing to a High degree of Self Conceit & obstinacy, that tho' they are certain, they are in an error, their pride wont allow them to own it & be put to rights again, this often proceeds from a Diffidence of heart too proud & too vain to See his error pointed out [VIII] but this should never be the Case, the Physician tho' ever should Conceal every such thing from his patients, or it might be to both their Prejudices, & the patient might Spoil the Physicians Character, & for fear of the danger might grow worse, but he should never Conceal the least error from his Brethren in the Faculty that they might put him to rights, again especially if they were men of Honour.

A Physician should have decorum, Decensy, & perspicuity of behaveour. –

I have already taken notice of the principle duties of a Physician to his patients. I shall now mention one more, w^{h} deserves notice, sometimes the patients, or their Relations will propose a Remidy, w^{h}. might do them good in their own opinion at least, Now the Physician often dispises all these, because for sooth they are not his prescription, this he should never do. but immediately give his opinion of it w^{t}. candour, & if he really aproves of it, tho' it did not before occur to him, he should by all means try it. but if he does not aprove of it let him say so, then if the patient or his friends still perist in applying it, Let them do so, why not let a man die in his own way if he will. – – A physi- [IX]

A Physician is sometimes at a loss what to say to Patients concerning their diseases & y^{e}. state of their health now in this case it would be very wrong to Acquaint the patient that he was really on y^{e}. point of Death, as this would hasten his death so much the sooner, now this may be a very important time for to acquaint his friends, as some minutes longer in life, might do a dale [deal] of service to y^{e}. family, therefore a lie in this case may be excusable in y^{e}. Physician to his patients, but he never ought to conceal the true Case from the Relations; this is the most Difficult part for [obscured, 'a'?] Physician, as sometimes they survive & then the Physician is ridiculed, & if he dies it striks the Relations more being foretold by the Physician. –

It has been alledged y^{t}. Physicians were all bad Irreligious men, but this is a mistaken notion, for many of y^{m}. [them] are good religious men, sometimes a Physician from y^{t}.circumstance alone may do more good to his patients, than all the drugs in the apothecarys shop, whereas a surly harsh manner does much harm to the Patients. –

Physicians in consultation should devest themselves of Particularities otherwise y^{e}. [they] sh^{d} [should] decline ye consultation altogether, in Consultations there should be a perfect argument among the Physicians otherwise if things mentioned in the Consultations as Se['crets be' ? part of page missing] spread abroad [X] abroad to the injuring or ridiculing of any one, in this Case Consultations do much more harm than good.

Antiently there was a Division of Medicine into three parts. 1st. The manual part or Surgery, 2^{d}: The Physicians part, & 3^{d}. The Apothecarys part.

In former ages Physicians applied themselves to the study of one or two Diseases only: but the Diseases of the Human body are so intimately Combined one w^{t}. another that is is equally impossible to understand Particulars without having a knowledge of the Whole, Every Disease both external & internal falls under the Cognisance of the Physician, Wherefore Surgery also belongs to ye Physician therefore Surgery & Physick should not be disjoined. –

A resolute Collected mind, a good Eye, and Steady hand, are the Qualifications requisite to make a good operator in Surgery, but the Qualifications of a Physician Differ widely f^{m} [from] this. – In some parts of Europe Surgeons act as Physician in ordinary to private families in other parts the appothecary do the same, the consequence of this is that the art is practised by a part of how illiterate men, thus in many parts [XI] parts of Europe Physicians of ye best Characters & Education, must depend all their life upon low mean apothecaries; w^{h} is very disagreable to y^{e}. Physician therefore the different parts of Physick sh^{d}: be differently professed or altogether by one man Sufficiently Educated for it.

A Physician is not a better man because of his <u>Degree</u>, a Degree gives a man no superior qualifications. – In many places Surgeons f^{m}. their Education act as Physicians in ordinary, & are sufficiently qualified for it too, as Gentlemen, as Physicians will easily find out these distinctions.

Much regard & Decorum has in all ages been payed to a certain formality in the Dress of Physicians w^{t}. a particular Deportment of Gravity, this may be of use in the Law, & other Professions where there is a necessity to strike the people w^{t}. awe & reverence, but no necessity for this in Physick, it is far f^{m}. supporting this Profession as has been supposed it did; it rather exposes the Professors to redicule, therefore I see no reason why a Physician sh^{d}. not dress like any other man, Let him agree w^{t}. the fashions of the Country he is in, if he is in a place where Physicians are distinguished by a laced coat a gold [XII] gold headed Cane & a large swinging sword, by his side, so that if he wanted these he would be less thought of, then by all means Let him wear them, but in general I see no need for a Physicians wearing a great Load of artificial hair, with a certain stiffness in his Coat & ce. [et cetera] –

A Physician sh^{d}. not have too much Delicacy, otherwise it will much disgust him about many circumstances in Practice, this Delicacy sh^{d}. on such occasions be laid aside, if the Physician sh^{d}. have to act as Surgeon, or even Nurse, when these are not at hand on certain occasions. – He should have Civility of manners, and no Falttery, w^{h}. is now so often made use of, he should be possessed of a very independent Spirit, and great ambition of mind, he sh^{d}. have no pride unless a certain degree accompanied w^{t}. other good qualities; he sh^{d}. have ease Gentleness, agreeableness of manners, & this is only to be attained by Practice.

A Physician sh^{d} not keep Nostrums or Misteries, w^{h}. is a great detriment to y^{e}. improvement of y^{e}. art, & as Nostrums w^{n}. devulged & sold for a triffle, its qualities (if it has any) are all lost immediately & in a short time it intirely vanishes & is quite forgotten, these Nostrums upon the whole do much more harm than good, they impose upon [fragment of page missing, probably: 'y^{e}'] judgement of the Physician & confound the Science [XIII] Science w^{t} false facts, & in this manner prevent the improvement of the Science. –

Physicians have been thought careless and Despisers of Religion, but this is not the case, for those Physicians who have made the greatest figures in y^{e}. world, have been marked out as good, Religious men, such as D^{r}. [Doctor] Boerhaave, Stahl[5], Sydenham[1], & cet. [et ceteri]. Religion is natural to the human mind, a Young Physician is very oft liable to be led aside from Religion thro pride, but very few Physicians have attempted to Redicule ye Principles of Religion, there are some particulars in y^{e}. Practice of Physick very well fitted to engage the heart of the Physician to Religion, from the many afflicting Schenes ['scenes'?] that so often necessarily comes under his eye enough to soften any heart.

I shall now proceed to mention, & to explain the Connexion of the several different branches of Physick with the practice of it, & how far it is necessary to have a knowledge of y^{e}. rest, in order to be qualified for ye practice of Physick there are a variety of little circumstances to be observed seemingly little connected together, everyone of these different branches are of infinite extent, but a student must [XIV] confine his views within certain bounds, and not at all pursue them to their extent, as there is no use for it, there are some however whose Genius leads them to the study of one particular branch more than of any other, thus we find some who are great Botanists, others great anatomists, & so on, who whoever ['however'?] makes but very poor & bad Physicians. – There is a necessity then for a knowledge of the other branches to a certain degree, thus the absolute necessity of a previous knowledge of anatomy, for y^{e}. practice of Physick is so aparently that it does not need mentioning. –

Next in order to understand y^{e}. animal œconomy, Physiology is necessary, then in order to understand y^{e}. different movements of the solid parts, of the œconomy, there is a necessity to be made acquainted w^{t}. the principles of Mechanics, & then for understanding the motion of our different fluids the knowledge of the principles of Hydraulics is necessary. – –

And next to comprehend y^{e}. nature & properties of the eye, there is a necessity of understanding the principle of optics, then as y^{e}. human body is surrounded w^{t}. a heavy dense fluid the Air, it is necessary to be acquainated w^{t}. its nature & properties, it is proper to be acquainted with the different [XV] different branches of Phylosophy for the sake of the animal œconomy both in its sound & disordered state, a good dale [deal] of y^{e}. nature & laws of y^{e}. animal œconomy, depend & are explained upon Chymical principles, hence y^{e}. necessity of the studying Chymistry previous to the Practice of Physick. –

The laws of y^{e}. animal œconomy are of the most different investigation. –

In order to attain to a true knowledge of human Physiology a knowledge of y^{e}. comparative anatomy is necessary, as many things have been found out of great use & advantage to Physick f^{m}. experiments made on Brutes.

The instinct of brute animals has given us great light w^{t}. regard to Diet, & other circumstances. It is impossible to discribe y^{e}. different Constitutions of Mankind in particular, but in general the difference of Constitution depends on the difference of age, climate, sex, Temperment & cet. this short detail shows us what, & how extensive a study Physick is. –

Physiology delivers the sound state of the body

Pathology the morbid. –

Materia Medica likewise is necessary for the Practice [XVI] Practice of Physick, as this gives the matter on which the Physicians works. –

A knowledge of mechanics is also necessary for the Practice of Surgery as y^{e}. surgeon works in a mechanical manner. –

A knowledge of Pharmacy is also necessary & a pretty extensive knowledge of Mechanics, Chymistry, & Physiology, are absolutely necessary as the effects of medicines on y^{e}. Human body are explicable sometimes on Mechanical, and sometimes on Chymical principles. –

Botany also is absolutely necessary to be known, as ye Physician sh^{d}. be acquainted with plants, w^{h}. he makes use of in Diet, & Medicine. I have now explained how far the other branhces of medicine are subservient to y^{e}. Practise of Physick. –

In Medicine the necessity of being acquainted w^{t}. y^{e}. principles of y^{e}. art, is much greater than in any other art. –

There is no necessity for a minute knowledge of y^{e}. Different Sciences, neither time for it, it is therefore necessary that a student should always have an Eye to the particular application of these to his Profession. – A Physi [XVII]

A Physician who has properly applied these must have an infinite advantage over those who are ignorant, of them there are some Physicians indeed who tho they have got no Education yet Practise w^{t}. Success being blissed [blessed?] w^{t}. a Happy natural Genius, while others of very Extensive learning, but wanting this natural Genius do not apply y^{r}. [their] judgement right & therefore make great blunders. –

There are other branches of learning a Physician sh^{d}. gett acquainted w^{t}. previous to Physick. –

I hope I have no occasion to insist on y^{e}. necessity of a knowledge of y^{e}. Latine Language, you sh^{d}. not however have too great an anxiety about Classical purity in your writings, as many authors by so doing miss what they chiefly aimed at, w^{h}. was by their eligance of stile to be better understood w^{n}. in reality they are worst. –

It is sufficient therefore to communicate our thoughts w^{t}. clearness, exactness, & simplicity, Celsus is a Standard of purity, & Eligance of learning, I must ['not' crossed out] also recommend y^{e}. study of that capacious eligant & expressive Language the greek then y^{e}. necessity of a knowledge of y^{e}. French Language is very apparent, as almost all the french authors who [XVIII] who are not few in number, write in their own language & also because now a days no Gentleman goes abroad without it. –

It is alleged that Physicians sh^{d}. hot have the accomplishment of other Gentlemen but in general I see no reason why a Physician sh^{d}. be excluded f^{m}. every or any accomplishment of a Gentleman as it renders him more easy & agreeable in Conversation, & affords an amusement & Relaxation of y^{e}. mind f^{m}. too much study. –

The General views to be attended to in the study of Nature, is first the advantage arising to individuals or 2dly. to y^{e}. public in general, 1st. To individuals it may serve as an amusement, or it may profite them, 2dly. as to publick utility it gives rise to all ye arts, it is favourable to ye interest of Religion by giving us a view of y^{e}. exact order and Regularity of every thing in Nature, it naturaly inspires us w^{t}. Love & awe for the Creator it has been alleged that a very extensive knowledge led to Atheism, but I rather imagine that is has quite the different effect, indeed to a silly minded man a little learning m[a]y do hurt, but not otherwise, Witness Lord Bacon, Mr Boyl, & Sir Isac Newton great men in this part of y^{e}. World but also good men. Now [XIX]

Now in all the different branches of natural knowledge, there is none more usefull & intertaining than that belonging to the Human Species. –

1st. Therefore the art of Medicine. –

2^{d}: The art of making People endure Cold, hunger, thirst, fatique, & cet. –

3^{d}: The art of improveing the different senses, both external & internal. –

These then are some of the most importent articles contained under the natural history of the human Species. –

Most men in point of knowledge have only views of gratifying their curiosity & of supporting, & profiting themselves, they have no views to y^{e}. publick utility, the Profession of Physick has suffered much in this way, and Physicians have corrupted this art by their own wild Extravagances. – I propose to sett down certain general rules for the practice of Physick, and to have no text book.

Nothing in nature falls out accidentally but every event happens in consequence of an universal law.

A principle of veracity is natural to the human mind, originally, hence you see in Children that they commonly tell the truth. We [XX] We obtain experience first by the evidence of our own senses 2dly. by the testimonies of others. – In every part of natural history and medicine above all others they have attended to facts only formed in their own imaginations, Medical writers Commonly abuse all other theories & writings but their own. –

Boerhaave, Hofman, & Stahl ['exclaim against' crossed out] use much theories, even Sydenham is full of hypothetical reasonings.

Those Physicians who Collect & arange all the facts relating to their Subject, & then bring out a direct proof by experiments, & if this fails that they do not stop here but being Sensible y^{t}. their is a fault somewhere, they endeavour to find it out, those I call rational Practising Physicians. –

Nothing can be more absurd than a Physician describing a Disease w^{n}. he in reality knows nothing about its cause. In our inquiry into Nature, a natural love of simplicity makes the Laws of Nature look fewer & more Simple, than they really are, Sir Isaac Newton Demonstrated many Phænomina of y^{e}. Simple Laws of Gravity w^{t}. so much clearness, & exactness, that it led them to a knowledge of many things [XXI] things that before they were ignorant of, he was led by analogy to believe y^{t}. all the Phænomina of y^{e}. Natural World depended upon Attracting & repelling powers, but this is not the Case, as there are powers evidently of a different nature in animal, & vegitable Substances. –

The different natures & Dispositions of men makes them different in their literary Charrecter thus an exuberancy of Genius makes men too restless & unsettled in y^{r}. notions of things, never bringing any thing to perfection, There are another sett of Geniuss w^{h}. are Calm & sedate, thus one man may possess both, he may have a Clear acurate Judgement, & at the same time a Calm Sedate mind, this Genius constitutes one of y^{e}. first Physicians in y^{e}. World, other again who are painfull, & industrious compilers, their Labours may be highly usefull. Some men particularly are extravagantly attached to prodigies, than to common appearances, & those are always remarkably too credulous. –

Natural History is more regarded as Curious Philosophy than as necessary for the practice of Physick or otherwise. – The [XXII]

The Study of Metaphysicks often does much harm by giving inginuety & industry a Wrong turn & derection. –

Diseases may be Classed acording to y^{r}. Symptoms proximate cause &ce [et cetera] all these ways are improper yet each of y^{m}. have advantages in y^{r}. own ways. much good has been lost in jangling about the arangement of Diseases.

It is of little importance to settle y^{e}. genera & Species of plants in Stead of Comparing their Virtues, y^{s}. [this] especially related to y^{e}. Practice of Physick. Dr. L[innaeus][6] has explained & reduced all the materials in nature, & treated y^{m}. in y^{e}. most perfect order; Diseases has been treated on the same plan w^{t}out [without] success. –

The advancement of the Science has been much hurt by a weak credulity of its Professors, a bigotted attachment to great names in the learned World, has done a dale [deal] of mischief to Physick by too great a regard to antiquity whatever is asserted to be a fact, deserve to be recorded till further illusterations.

A Blind Superstitious veneration for antiquity is another obstacle to the improvement of Science this retards the progress of knowledge while all other arts were improved Natural Phylosophy [XXIII] Phylosophy, & natural History lay intirely neglected till about y^{e}. middle of y^{e}. last Century Hipocrates was always esteemed in high veneration for his acurate & plain discription of Diseases, Galen commented upon His manner of writing Commentaries is absurd Hipocrates has left us a number of excellent observations, some of w^{h}. however true only in certain Cases, other false & absurd, his Commentator takes no trouble to find out Wether they are true or false, not only his observations but his opinions were often false, so that f^{m}. y^{e}. adoration Hippocrates there w[?]ins in a Physician thought himself under a necessity to show himself of y^{e}. same opinion, and what is worse even a Galen & in y^{s}. manner did mischief.

Nostrums & Quack medicines, the publick believed t^{t}. they were infallable cures in every Disease however they soon failed & then were laid intirely aside, because y^{e}. people thought that if they could not cure any Disease they could be of no use. –

Another obstacle to y^{e}. progress of y^{e}. science is y^{e}. Elegance of stile w^{h}. many authors attempt this Science sh^{d}. be delivered with all Simplicity & clearness possible the use of technical terms is now quite laid aside & all other absurdities of that nature. – I [XXIV]

I would recommend to your perusal y^{e}. Writings of Lord Bacon. –

Having thus endeavoured to explain some of y^{e}. principle causes obstructing y^{e}. improvement of the practice of Physick, I shall now mention some of y^{e}. peculiar disadvantages under w^{h}. a Physician Labours. –

1$^{\underline{st}}$. The general method of Conducting education in Coledges is not so well Calculated to improve Students in y^{e}. Science as to spread the knowledge of y^{e}. Science among many, the natural & general method of advancing Sciences is to advance general facts, in a particular order, y^{e}. young student thinks he understand every thing well, the method of curing every Disease so that little improvement is to be had f^{m}. him, as he thinks it does not stand in need of any, & if a patient dies he thinks he has done every thing for him, he should have done, or art could do, & being atached to certain opinions, it is very hard to get over y^{m}..

There are many Diseases w^{h}. no Physician has yet been able to cure, yet these Disease are not to be called incureable Diseases therefore no disease are to be incureable, but you can only go the length to say you dont know how to cure them to call a Disease uncureable is only a Cloak to Ignorance laizyness & slothfullness. Physici [XXV]

Physicians are neither better nor worse than the rest of mankind. –

A Physician w^{n}. he setts out in y^{e}. world Endeavours to gain ye applause of y^{e}. People, by difference in dress, by his gravity, & cet. he endeavours to put out his old Bretheren, but all his Receipts will remain in y^{e}. appothecarys file, to rise in judgement against him, no Considerable improvement can be expected f^{m}. Physicians on y^{s}. footing, Quacks have the advantage here, as they seldon [seldom] can suffer much either in their Charecter or practise, as a Quack has always more practise than a Regular Physician, they prescribe for more Diseases in a week than a Regular Physician does in a year, all the Quacks discriptions of Diseases are full of lies & absurdities. –

Medicine now begins to see a little more light than formerly it has done. –

No Science can boast of being more learned & ingenious than y^{t}. of Physick, this more than others gives a field for improvement.

A private Gentleman of a literary turn sometimes chooses to study medicine without any view of supporting himself. all he wants by it, is only to know y^{e}. goodness of those Books wrote in y^{e}. art & also w^{h}. is much more material to know [XXVI] to know, the learnedness & experience of those Physicians who take care of his health, w^{n}. in a disease & also those of his intimate friends.

A Physicians next to his own experience will learn most from y^{e}. Conversations of an ingenious & Candid Physician. –

AUTHORS

See D^{r}. Hales[7] one of the greatest benefactors of Physick, D^{r}. Cornai [Cornaro[8]?] wrote a Candid history of Diet, hence it will appear that I have no Opinion of Lessening the Importance of so great a Science. –
The true dignity of Physick is to be supported by Gentlemen of Candour & Equity. –

My business here then & y^{e}. nature of my Profession is to give a history of the General & particular Diseases of y^{e}. Human body, & a method how to cure the same. –

It is not my business here to explain Pathology or Therapeutics these belong to another Professor, In order to go through our Subject properly then it is necessary first to explain some General Division of Diseases this gives me an oppertunity of explaining many turns occuring in Physick, & first the Definition of a Disease, Every [XXVII] Every animal has certain functions attached to it according to its Sphere in life, now when it performs all those functions & actions aright in a natural easy way, then that animal is said to be in a Sound State of Body or in health, & every deviation f^{m}. this is Called a Disease so that in othe['r' is missing] words a Disease is every Deviation of the body from its natural state. –

All Definitions should consist as much as possible in plain certain facts. –

Some authors take in y^{e}. Cause in y^{e}. Definition, now that in general sh^{d}. not be y^{e}. Case.

The Definition of a Disease sh^{d}. consist in, & ought always to be taken from y^{e}. Symptoms, & always from these Symptoms w^{h}. are y^{e}. most constant.

In Some Cases however y^{e}. Cause of y^{e}. Disease must enter in to y^{e}. Definition but this is very seldom the case. – Definition [XXVIII]
Definition of a Disease
A Disease consists in any unnatural Structure or disordered function of the organs of the Body.
[I]

JOHN GREGORY

NOTES TAKEN FROM DR GREGORY

Citation:
Gregory, James [should be John]
Notes taken from Dr Gregory
reporter unknown
Royal College of Surgeons, Edinburgh: Ms. D 27, 1769.

Comment: The text contains contractions, wich are transcribed after the first occurrence of each. The ampersand is included where is appears in the text. Page numbers appear in brackets at the end of each page of original text. The text is numbered in Arabic numerals through p. 9. Emphasis is original throughout. Contractions are transcribed at their first occurrence. Original spellings are retained. Original underlinings are also retained; they appear for emphasis.

Acknowledgment: This documents appears with the permission of the Royal College of Surgeons of Edinburgh, I.B. Macleod, Honorary Secretary. I am grateful to the Librarian of the College, Miss A.M. Stevenson, for her corrections to my initial transcription.

Text:

Notes taken from Dr Gregory

Nov^{r}. [November] 6th 1769 I am this day to mention to you some of the Moral Qualities of a Physician & ce. [et cetera] And the first I shall mention is Humanity or that Compassion we ought to have for our Patient. Some have alleged that our Profession are very hard hearted & void of feelings. but I hope the Charge is groundless (with respect to the generallity of Phys^{ns}. [Physicians][)] Real Sympathy is of the greatest advantage both to the Pat. [Patient] & Physician: real sympathy always trys to conceal itself – Another qualification

I shall [1] mention, is a certain ease & Gentleness to be shewn towards the Pat. & not to be too rigid & stiff w^t [with] regard to particulars which are sometimes not of the greatest Consequence – We should not e.g. be too severe in recommending to ['y' crossed out]our Pat. a Particular Regimen, which is very often not observed. I have very often observed that where the Phy^n. was scrupulously strick [strict] is [in?] this point, that they who attended the Sick, were carefull in concealing the transgression from the Phys^n. We should never prescribe any Rule where there is no probability of its being observed. [2] yet I would not have a Physician to give Sanction to any thing that might be hurtfull to the Pat. but in the gentlest manner to endeavour to persuade him by the use of it –

Another thing I would mention to the Phys^n. – that in Nervous Disorders, when the Pat. is pivish & discontented & has a great many complaints which may appear very ridiculous, in this case whether are we to trait [treat] them w^t negigence or Ridicule? we are to do no one of them but to endeavour to strike a proper medium betwixt them: but to [3] to endeavour to divert the Pat^s. attention from his complaints, for I think it is as much the Business of the Phys^n. to endeavour to remove the diseases of the Mind as those of the Body. they being as real as those that evidently appear to us. The treating them w^t contempt or Ridicule argues a shallow & weak mind –

The Phys^n ought also to be <u>candid</u> & allow himself to be convicted – The man who thro' pride, (or which is more commonly the case) ignorance, will not allow himself to be convicted when he may be in a mistake, does great dishonour to y^e [the] profession. [4]

When two Physicians are consulting the case of the same Pat. they ought to lay aside all private malice & resentment at that time, & w^t the candour & free spirits of Gentlemen, openly & w^t honour express their sentiments. One who cannot keep his temper w^t [something crossed out; not legible; 'one'?] a Phys^n. he has any quarl with, ought to refuse the Consultation, because their determinations, if any, might be of very little use to the Pat. –

I shall now make some remarks w^t regard to the <u>Decorum</u> necessary to a Phys^n.. With respect to his behaviour & deportment, I think it ought to be the same w^t that of any other Gentleman, his be [5] haviour ought to be easy & affable with respect to the Dress if fashion of the Country he lives in require any alteration from that of ['a' crossed out] private Gentlemen, we ought to conform to that Custom – but when there is no need of alteration, it argues great weakness to affect solemnity or Distinction in this respect

There is one thing that particularly requires your attention, & that is the telling a Pat. the true state & hazard he is in when labouring under any Disease & here a small Deviation from truth is sometimes requisite [6] He who comes in to a Pat. & in a harsh & brutal manner tells him he is dying, is acting the part of one giving the sentence of death rather than a Physician

Yet I would be loath to conceal from any person the state he was in, if any thing of great consequence depended upon him, as the ruin of his family, the settling of his affairs & ce.

Again, you ought never to leave your Pat. altho' you plainly see you can give him no relief but endeavour as far as is in your power to alleviate the Symptoms of his Disease: even the presence [7] of a Phys^{n} has great effect in easing the mind of the Pat.

Our profession have many opportunities to [a letter scratched out; not legible] see the private affairs and actions of those they attend & particular among the women: A Phys^{n} who takes the advantage of them, I mean one that is amorous is highly culpable.

A Phys^{n}. ought to be particularly acquainted w^{t} Chyrurgery, even altho' He does not practise it

As to the keeping secrets or nostrums this is a great disadvantage to the society. The argument they use is that if it were known it would perhaps not be much [8] regarded but the Contrary appears to me to be true – In general nostrums do a great deal more hurt than good, as they confound the judgement w^{t} false facts –

The Profession hath also been branded w^{t} Infidelity & contempt of Religion. Yet I would flatter myself that the Charge is groundless; the reason is those who have got a liberal Education, are above being Bigots to any kind of Religion. The Phys^{n}. (if he can w^{t} propriety) should advise them to get a Clergyman, for he may do more good in composing the the Pat.^{s} mind than the skill of any Physician [9]

JOHN GREGORY

NOTES OF LECTURES ON THE PRACTICE OF PHYSIC

Citation:
Notes of Lectures on the Practice of Physic, begun 2 November 1772 by H. Morris.
Royal College of Physicians and Surgeons, Glasgow: MS 1/9/10, 1772.

Comment: The text contains very few contractions. The ampersand is included where it appears. Page numbers appear in brackets at the end of the page of the original text. There is an insert to p. 4 on p. 3v (verso). The text is numbered in Arabic numerals through p. 9. Page 10 begins discussion of "Febrile Diseases." Original spellings are retained.

Acknowledgment: This document appears with the permission of the Library and Archive Committee of the Royal College of Physicians and Surgeons of Glasgow, James Beaton, Archivist and Librarian.

Text:

Notes taken from D^{r}. [Doctor] Gregory's Lectures on the Practice of Physic begun 2^{d}. Nov^{r}. [November] 1772.
The art of medicine seems to be coevial with man, and at first to have taken its rise from the natural instinct implanted in men as well as other animals, from medicines tried at random in order to give ease in pain or sickness and from observing the methods nature took to cure diseases – hence we cannot seek its origin from any particular country – Asculapius the son of Apollo is celebrated as the inventor of Physic, and his two sons are said to have gone to the siege of Troy after this death he was wirshoped as [1] a god, to his temple the sick came and were cured, all of the cures were ingraved upon tablets and hung up in the temple, the art of medicine was preserved in this family for 700 years, from hence Hipocrates sprung who lives about 400 years before christs, he freed physic of philosophy at least he join'd faithfull observation and practice, he is the

oldest writer upon this art, the writings which seem to be truely his are truely judicious and simple he is the first that strictly observed the name of Physician, there are several books that pass under his name which seem to be spurious, his prognostics and Aphorisms contain many things valuable [2] and usefull, he lived till he was 104 years of age. For many ages after Hipocrates physic seems to have been little improved ['the' crossed out]his followers were called Empirics, they disregarded reasonings & remote causes trusting to observation & experience alone – Themothon[9] was founder of the methodic set [sect] who imputed all diseases either to Stricture, Laxity or a mixture of the two and accordingly they directed their cures – afterwards lived Galen he came to Rome where he practised with great reputation he confuted all the then rei['gh' crossed out]ning sets [sects] and pretended to revise Hipocrates, he wrote a vast deal and was truely a man of genius and of great learning but did very much harm to the medical art by his wild Theories, than he ever did service, his doctrine was the only one that prevailed for about 1500 years during all this period [3] there was little improvement made in physic there w['as' crossed out]ere many books wrote particularly by the Arabians who preserved the Greek & Roman learning when these nations were overrun by the barbarious northern nations but all of them compiled from or commented upon Galen – About the latter end of the the 15.th century Paracelsus[10] was born in Switzerland, he had a regular education for a physician, but being of a bold & fertile genius he travelled thro' different countries in search of chemical remedies & the Phylosopher's stone, he went thro' as an itinerent boldly prescribing his chemical remedies & sometimes with such success that made him boast of being possest of a remedie that would prolong life beyond its ususal limits, he cried out against Galen and the Antients and was the first that A introduced chemistry into medicine ['A' refers to the following passage at the bottom of p. 3r: Paraselsus made great use of Mercury, Antimony and Opium. the former which he gave in the preparation of red proecip. gain'd him great carecter in the veneral dissease which made its appearance about this time baffeling ye efforts of ye Galenists] – The next [4] [p. 5 is unnumbered and blank; the next numbered page of text is p. 6.] considerable revolution that took place in medicine was by Dr Harvey's[11] celebrated discovery of the circulation of the blood which overturned all the Theories which so many former years had established. and built medicine upon a more firm foundation that it had ever before stood ['upon' crossed out] Dr Harvey's book was first published in the year 1628 not without much oppossition – The chemical Physicians had still been gaining ground over the Galenical ever since Paracelsus, they ascribed diseases to acidity, formentation & ca [et cetera], sometime after Harvey the mathematics began to be studied with great eagerness, this led Baglivi[12] first to consider the body in a mechanical view by which he endeavoured to account for its different diseases upon mechanical principles as an inanimate machine, he was followed by Bellini[13] and others, hence were introduced laxity & rigidity sentor & ca – Physic thus stood when Boerhaave [6] to whom medicine is much indebted found it, he confined himself to no particular sect, his great loss was in considering the body too mechanically and not giving the mind its due share – Different from him was Stahl[5], he ascribed all to the mind and made all the motions both voluntary & involuntary immediately under its power and even life itself; he declaimed against opium, astringents, the Bark & ca as disturbers of nature

After giving the general history of Physic he told in what order he is to proceed in his descriptions of diseases; first to give an account of the antecedent symptoms or those appearances which precede the disease. 2.dly The diagnostic symptoms or those serving to distinguish it from others 3.dly The causes as far as can be learn'd either from the predisposing causes or those which render y.e bodie susceptable of a disease by means of an occational cause tho' sometimes of themselves [7] sufficient to produce the disease in this they differ from seminia which can never produce disease without an occational cause, as in the smallpox the seminium never produces the disease without infection – The occational cause is what acts immediately upon the system in producing disease such as cold iregularity in diet & ca these two causes togcither are called remote – The proximate cause is the concurance of all the causes necesary to produce the disease or from which it immediately proceeds – as in an asthma the remote cause may be too great irritability in the lungs the occational flatulance in the stomach & the proximate is the constriction of the vessels of the lungs.

Diseases have been classed upon different principles from predisponent & from occational causes those that affect a particular part have [8] been class'd togcither as those which affect the head all these different ways have no certainty the best way would seem to be by their proximate causes but the uncertainty of these in many cases renders this impossible to be done with certainty – the way most commonly at present attempted & indeed the most practicable is by the symptoms – They are divided into classes, orders, genera, species and varieties – The definition of a disease ought to include the particular Diagnostic symptoms which always accompanie it & serve to distinguish it from every other disease – it ought not to include the cause unless in some few instances – nor the continuance, in some cases the symptoms that appear at any particular stage are not sufficient to distinguish it as in the small Pox and other eruptive diseases, the description ought therefore to comprehend this as well as the Febrile symptoms. [9]

JOHN GREGORY

OBSERVATIONS ON THE DUTIES AND OFFICES OF A PHYSICIAN AND ON THE METHOD OF PROSECUTING ENQUIRIES IN PHILOSOPHY

Citation: Gregory, John
Observations on the Duties and Offices of a Physician and on the Method of Prosecuting Enquiries in Philosophy
London: Printed for W. Strahan; and T. Cadell (Successor to Mr Millar) in the Strand
1770

Comment: No changes have been made to the original text. In particular, original spellings and punctuation are retained. Spellings of particular words may vary in the text; these variations have been retained. The goal has been to replicate the original text in its entirety. Page numbers appear in brackets at the end of each original page number. '[n.p]' is used to indicate a page break in the original to unnumbered pages.

Acknowledgment: This documents was prepared from the copy of *Observations* in the Blocker History of Medicine Collections in the Moody Medical Library of the University of Texas Medical Branch, Galveston, Texas. I am grateful to Inci Bowman, Ph.D., Curator of the Blocker Collections for her assistance.

Text:

OBSERVATIONS
ON THE
DUTIES & OFFICES
OF A
PHYSICIAN
AND ON THE
METHOD OF PROSECUTING
ENQUIRIES IN PHILOSOPHY

London
Printed for W. Strahan, and T. Cadell
(Successor to Mr. Millar) in the Strand
MDCLXX [n.p.]

Advertisement

The following sheets contain two preliminary lectures, read not long ago, in one of the universities of a neighbouring kingdom, by a medical professor.

Many copies, from a general satisfaction they afforded his audience, were taken down in short-hand. Of these the reader is here presented with the most correct; and the editor flatters himself, that from the free and liberal spirit of enquiry which animates the whole of them, they will provide a most acceptable present to the public; and, of course, do no discredit to the ingenious author.

In the first lecture, the author has treated fully on the duties and of-[iii]fices of a physician: a path almost untrod till now. The noble and generous sentiments which are here displayed, will ever be a source of pleasure to minds unbiased by prejudice, self-interest, or the unworthy arts of a Corporation. Whatever opposition this part of the work may meet with from those, who find their own foibles, or rather vices, censured with a just severity, the ingenious part of mankind, however, will not fail in bestowing that degree of applause so justly due to its merit. At present there seems to be a general disposition in mankind to expose to their deserved contempt, those quackish, low, and illiberal artifices, which have too long disgraced the profession of medicine. It is therefore hoped, that the general spirit of this lecture will have a remarkable tendency to promote this laudable end; and that it will excite men of influence and abilities to exert themselves in crushing that arrogance, which hath frequently served to cover the ignorance of many practitioners of medicine, and [iv] by means of which alone, they acquire such a share of practice as they are by no means entitled to. In consequence of this, real merit, which is very often accompanied with great modesty, will meet with its due reward.

"Vir bonus & sapiens dignis ait esse
"paratum:
"Nec tamen ignorat quid distent aera
"lupinis." HORACE

Those physicians, on the other hand, who are men of erudition and genius, and have the mens sibi conscia recti, and who are possessed of those amiable virtues, which are the ornaments of human nature, have nothing to fear from anything advanced in the following pages: on the contrary, they will find, that every good quality they possess hath an obvious tendency, not only to advance the dignity of the science which they profess, but also to promote their own interest and reputation. Students of medicine will also reap much advantage from this part of the present pub-[v]lication; since they will see clearly, what course they ought to pursue, in order to attain the ultimate object of their desires, and, at the same time, will discover the rocks, upon which so many of the profession have been wrecked.

In the second lecture, the author has endeavoured to ascertain the true method by which enquiries into medicine are to be prosecuted; and has likewise pointed out, with much precision, the causes which have retarded the advancement of medicine, and the inconveniences which that science in particular at present labours under.

Lord Bacon justly observes that medicine can never be improved till its imperfections are pointed out. Our author, therefore, hath attempted to explain the leading principles, according to which medical enquiries ought to be conducted. In order to this, he considers medicine as a branch of natural philosophy, to be prosecuted with the same general views, and upon the same [vi] plan. The observations have therefore a reference to this subject in general; and when necessary, a particular application is made to the practice of physic. Enlarged views of nature, of the connection of the sciences, the method of advancing them, and the causes which have hitherto retarded their advancement, render a person better qualified to study that particular department of science, to which he chuses to attach himself.

In the prosecution of this subject, the author has had occasion to adopt many sentiments of Lord Bacon, as delivered by that incomparable judge of philosophical writing, in his books De Augmentis Scientiarum & Novum Organum; which, however, are illustrated by examples more particularly taken from the history of physic. This part of the work, the editor apprehends, will be of remarkable utility to those who are entering upon the study of medicine, by directing them to the proper fountains from whence true medical knowledge can [vii] be drawn, and by teaching them to distinguish between what is true and what is false; wherein a well-founded theory consists; and how far it differs from that scholastic jargon of metaphysical chimeras, and new-coined unmeaning words, so frequently used by medical writers. This is a most important object to every person who studies medicine, either with a view to his own improvement, or to the farther advancement of the science.

Had the author himself published these lectures, they would doubtless have appeared to greater advantage. More accuracy, with regard to composition, would certainly have been displayed, and many judicious observations would probably have been added, tending to elucidate several particulars of the extensive subject of which he treats. But notwithstanding this, the editor hopes, that the language is sufficiently correct, and that the meaning of the author is every where delivered with precision and perspecuity. [viii]

LECTURE I

The design of the office which I have the honour to hold in this university, is to explain the practice of physick; by which I understand, the art of preserving health, of prolonging life, and of curing diseases. An art of great extent and importance; for which all your former medical studies were intended to qualify you: and indeed it is only so far as they tend to qualify you for this, that they are of consequence to you as physicians.

Before I proceed to the particular business of this course, I shall, agreeably to the usual custom, give some preliminary lectures. Such lectures are intended [1] to have a relation to the proper subject of the profession, but not to be essentially connected with it. – On this occasion I think it needless to dwell on the utility and dignity of the medical art. Its utility was never seriously called in question; every man who suffers pain or sickness will very gratefully acknowledge the usefulness of an art which gives him relief. People may dispute, whether physick, on the whole, does most good or harm to mankind? just as they may dispute, whether the faculty of reason, considering how it is often perverted, really contributes to make human life more or less happy? whether a vigorous constitution and an independent fortune are blessings or curses to those who possess them? whether the arts and sciences in general have proved beneficial or detrimental to mankind?– Such questions afford opportunities for the display of eloquence, and for saying plausible and ingenious things; but still nobody doubts of the real and substantial [2] advantages attending those acquisitions, if the natural and proper uses are made of them. Much wit has indeed, in all ages, been exerted upon our profession; but if we attend to it, we shall find that this ridicule has rather been employed against physicians than physick. There are some reasons for this sufficiently obvious. Physicians, considered as a body of men, who live by medicine as a profession, have an interest separate and distinct from that of the art. In pursuit of this interest, some have acted with candour, with honour, with the ingenuous and liberal manners of gentlemen. Conscious of their own worth, they disdained all artificial colourings, and depended for success on their real merit. But such men are not the most numerous in any profession. Some impelled by necessity, some stimulated by vanity, and others anxious to conceal conscious ignorance, have had recourse to various mean and unworthy arts to raise their importance among the ignorant, who are always the most numerous part of mankind. Some of these [3] arts have been an affectation of mystery in all their writings and conversations relating to their profession; an affectation of profound knowledge, inscrutable to all, except the adepts in the science; an air of perfect confidence in their own skill and abilities; and a demeanour solemn, stately, and highly expressive of self-importance. These arts, however well they might succeed with the rest of mankind, could not escape the censure of the more judicious, nor elude the ridicule of men of wit and humour. Accordingly, it has been pointed out against them with so much keenness, that we never meet with a physician in a dramatick presentation, but he is treated as a solemn coxcomb and a fool. But it is evident, that all this satire is levelled against the particular manners of individuals, and not against the profession of medicine itself.

Of the dignity of the profession I need say little. I suppose you are well satisfied that you have chosen a cre-[4]ditable one. Whatever may have been the customs or caprices of a few particular countries, it has generally been looked upon, as one of the most liberal professions, and on the justest grounds. To excel in it requires a greater compass of learning than is necessary in any other art.– A knowledge of the mathematicks, at least of the elementary parts of them, of natural history, and natural philosophy, are essentially connected with it; as well as the sciences of anatomy, botany, and chemistry, which are deemed its immediate branches. There are likewise some pieces of knowledge, which, though not absolutely necessary to the successful practice of medicine, are yet such ornamental acquisitions, as no physician who has had a regular education is found without; such are, an acquaintance with the Latin, Greek, and French languages. If you add to this, that knowledge of the world, of men, and of manners, so useful to a physician, and which he naturally and insensibly imbibes from an [5] extensive intercourse with mankind, I think it will evidently appear, that no profession requires a greater variety of liberal accomplishments than that of physick; and this sufficiently establishes the dignity of the science.

We have particular reason to be pleased with the honourable point of view, in which our profession is regarded in every part of the British dominions. They who have seen the contemptible light in which some of its branches are considered in some other countries of Europe, will feel more sensibly the just regard paid to them in their own country. One happy consequence, among many others, which results from this, is, that gentlemen of the best families, distinguished for their spirit and genius, often apply to the study of medicine; and the liberal and ingenuous manners, naturally to be expected from men well born and genteely educated, reflects an additional dignity on the profession. [6]

Besides the general consideration of the utility and dignity of the science of medicine, it may be considered in two different views.

1. In the first place, as presenting a very ample field for the exertion of genius. – The great extent of the subject, and a variety of causes, which I shall afterwards endeavour particularly to explain, have left it imperfect in many of its parts; and indeed there are some paths in it hitherto untrod.

2. In the second place, medicine presents an equally extensive field for the exercise of humanity. A physician has numberless opportunities of giving that relief to distress, which the wealth of India could not purchase. This, to a benevolent mind, must be one of the greatest pleasures it

can ever hope to enjoy. But besides the good which a physician has it often in his power to do in consequence of skill in his profession, there are many occasions that call for his assistance as a [7] man, as a man who feels for the misfortunes of his fellow creatures. In this respect he has many opportunities of displaying patience, good nature, generosity, compassion, and all the gentler virtues that do honour to human nature. I do not here speak of the obligations of principle. I speak only of what nature dictates to a heart possessed of any portion of goodness and sensibility. Our faculty have often been reproached with hardness of heart, occasioned, as is supposed, but their being so much conversant with human misery. I hope and believe the charge is unjust. Habit may beget a command of temper, and a seeming composure which is often mistaken for absolute insensibility. When this insensibility is real, it is a great loss to a physician, as it deprives him of one of the most natural and powerful incitements to exert himself for the relief of his patient. On the other hand, a physician of too much sensibility may be rendered incapable of doing his duty by anxiety, and excess of sympathy, which [8] cloud his understanding, depress his spirit, and prevent him from acting with that steadiness and vigour, upon which perhaps the life of his patient in a great measure depends.

This naturally leads me to make some observations on the duties and office of a physician; a subject of great importance, but perhaps of too delicate a nature for a physician to treat of with openness and freedom. On this, however, I should be altogether silent, if I did not feel myself at full liberty to discuss it without any reserve. The difficulty of treating this subject in such a manner as to give no offence arises from hence, that medicine may be considered either as an art the most beneficial and important to mankind, or as a trade by which a considerable body of men gain their subsistence. These two views, though distinct, are far from being incompatible, though in fact they are too often made so. I shall endeavour, however, to set this matter in such a light as may shew that the system [9] of conduct in a physician, which tends most to the advancement of his art, is such as will most effectually maintain the true dignity and honour of the profession, and even promote the private interest of such of its members as are men of real capacity and merit. I am under less apprehension of discussing this subject before gentlemen at your period of life, than if you were further advanced in years. Youth indeed is the season when every sentiment of liberty, of generosity, and of candour, most easily find their way to the heart. If they do not reach it then, they never will afterwards. Age may improve the understanding by accessions of knowledge and experience; whilst at the same time that warmth of temper and imagination, which so often mislead the judgment, gradually subside. But it unfortunately happens, that this very circumstance, which in some respects improves the understanding, in others throws a damp upon genius, checks the ardent pursuit of science and truth, and [10] shuts the heart against every manly, enlarged, and generous sentiment.

In the prosecution of this subject, I shall, in the first place, consider what kind of genius, understanding, and temper naturally fit a man for being a physician.– In the second place, what are the moral qualities to be expected from him in the exercise of his profession, viz. the obligations of humanity, patience, attention, discretion, secrecy, and honour, which he lies under to his patients.– In the third place, I shall take notice of the decorums and attentions peculiarly incumbent on him as a physician, and which tend most effectually to support the dignity of the profession; as likewise the general propriety of his manners, his behaviour to his patients, to his brethren, to surgeons and apothecaries. In the fourth place, I shall particularly describe that course of education which is necessary for qualifying a physician to practise with success and reputation; and shall, at [11] the same time, mention those ornamental qualifications expected from the physician as a gentleman of a liberal education, and without which it is difficult to support the honour and rank of the profession. –

I. I begin with an enquiry into the genius, understanding, and temper, which naturally fit a man for being a physician.

Perhaps no profession requires so comprehensive a mind as medicine. In the other learned professions, considered as sciences, there is a certain established standard, certain fixed laws and statutes, to which every question must constantly refer, and by which it must be determined. A knowledge of this established authority may be attained by assiduous application and a good memory. There is little room left for the display of genius, where invention cannot add, nor judgment improve; because the established laws, whether right or wrong, must be [12] submitted to. The only exercise for ingenuity, is in cases where it does not clearly appear what the laws are. But even then, as disputable points must be referred to the determination of judges, whose opinions, being formed from various circumstantial combinations, frequently differ, there is no criterion by which the ingenious reasoner can be judged; and his conclusions, whether well or ill drawn, must still remain undecided. The case is very different in medicine. There we have no established authority to which we can refer in doubtful cases. Every physician must rest on his own judgment, which appeals for its rectitude to nature and experience alone. Among the infinite variety of facts and theories with which his memory has been filled in the course of a liberal education, it is his business to make a judicious separation between those founded in nature and experience, and those which owe their birth to ignorance, fraud, or the capricious systems of a heated and [13] deluded imagination. He will likewise find it necessary to distinguish between important facts, and such as, though they may be founded in truth, are notwithstanding utterly useless to the main ends of his profession. Supposing all these difficulties got over, he will find it no easy matter to apply his knowledge to practice. In teaching a system of the practice of physick, every disease must be considered separately, and as existing by itself; but in fact diseases are found complicated in endless varieties, which no system, however perfect, could possibly admit. This occasions an embarrassment to a young practitioner, which nothing can remove but a habit of nice discrimination, a quickness of apprehension which enables him to perceive real analogies, and, what is rarely united with this, a solidity of judgment, which secures him from being deceived by imaginary ones. A student of much fancy and some learning, has no idea of this embarrassment. In the pride of his heart, he fancies [14] every disease must yield before him; he thinks he not only knows the proximate causes and indications of cure in every disease, but a variety of remedies that will exactly fulfil these indications. It will be unfortunate however for his patients, if a little experience does not humble his pride, and satisfy him that in many cases he neither knows the proximate causes nor the indications of cure, nor how to fulfil these indications when he does know them; or shew him, what is equally perplexing, that the indications are different and contradictory. In this situation his boasted science must stoop, perhaps, for some time, to be an idle spectator, or to palliate the violence of particular symptoms, or to proceed with the utmost caution and diffidence, on some very loose and precarious inductions from analogy. Such are the difficulties which a physician has to encounter in his early practice; to conquer which is required, independent of all the assistances of a proper education, the con-[15]currence of an acute, penetrating genius; and of a clear, solid judgement; and, in many cases, of a quickness of apprehension, which instantaneously perceives where the greatest probability of success lies, and seizes the happy moment of action.

But, although a physician should possess that enlarged medical genius, which I have just now described, a capacity of another kind is also required. A physician has not only his own prejudices to conquer, but he must study the temper, and struggle with the prejudices of his patient, of the relations, of his own brethren, of the world in general; he must guard himself against the ill offices of those, whose interests interfere with his; and it unfortunately happens, that the only judges of his medical merit, are those who have an interest in concealing or depreciating it. Hence appears the necessity of a physician's having a large share of common sagacity, and knowledge of the world, as well as a medical genius and erudition. [16]

Such is the genius and capacity required in a physician; but a certain state of the temper and passions, either natural or acquired, is requisite, in order to give them their full advantages. Sudden emergencies occur in practice, and diseases often take unexpected turns, which are apt to flutter the spirits of a man of lively parts and of a warm temper. This may embarrass his judgment in such a manner as to disable him from discerning what is proper to be done, or if he does discern it, it may render him irresolute in his conduct. Yet such occasions call for the quickest discernment and the steadiest and most resolute conduct. The follies and bad behaviour of patients, and a number of little difficulties and contradictions which every physician must encounter in his practice, are likewise apt to ruffle his temper, and consequently to impair his judgment, and make him forget the propriety and decency of his behaviour. Hence appears the necessity of a physi-[17]cian's possessing great presence of mind, composure, steadiness and resolution in acting, even in cases where, in his private judgment, he is extremely diffident. It is also necessary to acquire such a command of temper, as may enable him to conceal his diffidence or embarassment, both for the patient's sake and his own.

II. I come now to mention those moral qualities peculiarly required in the character of a physician. The most obvious of these is humanity; that sensibility of heart which makes us feel for the distresses of our fellow creatures, and which of consequence incites us in the most powerful manner to relieve them. Sympathy produces an anxious attention to a thousand little circumstances that may tend to relieve the patient; an attention which money can never purchase: hence the inexpressible advantages of having a friend for a physician. Sympathy naturally engages the affection and confidence of a patient, which, in many cases, is of [18] the utmost consequence to his recovery. If the physician possesses softness and gentleness of manners, a compassionate heart, and what

Shakespeare so emphatically calls "the milk of human kindness," a patient feels his approach like that of a guardian angel ministering to his relief; while every visit of a physician who is unfeeling, harsh or brutal in his manners, makes his heart sink within him, as at the presence of one, who comes to pronounce his sentence of death. Men of the most compassionate tempers, by being daily conversant with scenes of distress, acquire in process of time that composure and firmness of mind so necessary in the practice of physick. They can feel whatever is amiable in pity, without suffering it to enervate or unman them. Such physicians as are callous to every sentiment of humanity, affect to treat this sympathy with great ridicule, and represent it either as hypocrisy, or the indication of a feeble mind. That it is often affected is beyond [19] question; but this affectation is easily seen through. Real sympathy is never ostentatious; on the contrary, it rather strives to conceal itself. But, what most effectually detects this hypocrisy, is a physician's different manner of behaving to people in high and people in low life; to those who fee him genteely, and those who cannot fee him at all. A generous and elevated mind is even more shy in expressing sympathy with those of better rank, than with those in humbler life; being jealous of the unworthy construction so usually annexed to it. – The insinuation that a compassionate and feeling heart is the effect of a feeble mind, is equally replete with malignity and falsehood. Universal experience demonstrates, that a gentle and humane temper, so far from being inconsistent with vigour of mind, is its usual attendant; and that rough, blustering manners generally accompany a weak understanding and a dastardly soul, and are indeed frequently affected by men void of mag-[20]nanimity and personal courage, to conceal their natural infirmities.

There is a species of good nature different from the sympathy I have been speaking of, which is very amiable in a physician. It consists in a certain gentleness and flexibility, which makes him suffer with patience, and even apparent chearfulness, the many contradictions and disappointments he is subjected to in his practice. If he is extremely rigid and particular in his directions about regimen, he may be assured they will not be strictly followed; and if he be severe in his manners, the deviations from his rules will as certainly be concealed from him. The consequence is, that he is kept in ignorance of the true state of his patient; he ascribes to the consequences of the disease, what is merely owing to irregularities in diet, and attributes effects to medicines which were every day thrown out of the window. The dangerous errors which in this way he may [21] be led into, are sufficiently obvious, and might easily be prevented by a prudent relaxation of rules that could well never be obeyed. The government of a physician over his patient should undoubtedly be absolute but this absolute government very few patients will submit to. A prudent physician should, therefore, prescribe such laws, as, though not the best, are yet the best that will be obeyed; of different evils he should choose the least, and, at any rate, never lose the confidence of his patient, so as to be deceived as to his true situation. This indulgence, however, which I am pleading for, must be managed with great judgment and discretion; as it is very necessary that a physician should support a proper dignity and authority with his patients, for their sakes as well as his own. There is a numerous class of patients who put a physician's good-nature and patience to a severe trial; those I mean who suffer under nervous ailments. Although the fears of these patients are [22] generally groundless, yet their sufferings are real; and the disease is as much seated in the constitution as a rheumatism or a dropsy. To treat it with ridicule or neglect, from supposing it the effect of a crazy imagination, is equally cruel and absurd. It is generally produced or attended with bodily disorders, obvious enough; but supposing them not obvious, still it is the physician's duty to do every thing in his power for the patient's relief. Disorders of the imagination may be as properly the object of a physician's attention as a disorder of the body; and surely they are, frequently, of all distresses the most dreadful, and demand the most tender sympathy: but it requires great address and good sense in a physician to manage them properly. If he seems to treat them slightly, or with unseasonable ridicule, the patient is shocked beyond measure; if he is too anxiously attentive to every little circumstance, he feeds and rivets the disease. For the patient's sake therefore, as well as his [23] own, he must endeavour to strike the due medium between negligence and sarcastic ridicule on the one hand, and an anxious solicitude about every trifling symptom on the other. He may sometimes divert the mind, without seeming to intend it, from its present sufferings, and from its melancholy prospects of the future, by insensibly introducing subjects that are amusing or interesting; and sometimes he may successfully employ a delicate and good-natured ridicule. – It is not unusual to find physicians treating these complaints with the most barbarous neglect, or mortifying ridicule, when the patients can ill afford to see them; while at the same time, among patients of higher rank, they foster them with the utmost care and apparent sympathy: there being no diseases, in the stile of the trade, so lucrative as those of the nervous kind.

We sometimes see a remarkable difference between the behaviour of a [24] physician at his first setting out in life, and afterwards when he is fully established in reputation and practice. When beginning the world he is affable, polite, humane, and assiduously attentive to his patients: but when in the process of time when he has reaped the fruits of such a behaviour, and finds himself above the world, he assumes a very different tone; he becomes haughty, rapacious, careless, and sometimes perfectly brutal in his manners. Conscious of the ascendency he has acquired, he acts the part of a despotic tyrant, and insolently boasts, that no man, in the place where he resides, dare die without his leave. He not only takes a most ungenerous advantage of the confidence which people have in his abilities, but lives upon the effects of his former reputation, when all confidence in his abilities has ceased: because a physician who has once arrived at a very extensive practice, continues to be employed by many people for their friends, who think of him themselves [25] with contempt; they employ him because it is fashionable to do so, and because they are afraid, if they neglected it, their own characters may suffer in the world.

A physician, by the nature of his profession, has many opportunities of knowing the private characters and private transactions in families. Besides what he learns from his own observation, he is often admitted to the intimate confidence of those, who perhaps think they owe their life to his care. He sees people in the most disadvantageous circumstances, very different from those in which the world views them; – oppressed with pain, sickness, and low spirits. In these humbling situations, instead of usual chearfulness, evenness of temper, and vigour of mind, he meets with peevishness, impatience, and a spirit perfectly broken and ennervated. Hence it appears how much the characters of individuals, and the peace and happiness of [26] families, may sometimes depend on the discretion, secrecy, and honour of a physician; who, without deviating from truth, may render characters that are truly respectable, ridiculous and contemptible. The most profound secrecy is particularly requisite where women are concerned. Independent of the peculiar tenderness with which a woman's character should be treated, there are certain circumstances of health, which, tho' in no respect connected with her reputation, every woman, from the natural delicacy of her sex, is extremely anxious to conceal; and, in some cases, the concealment of these circumstances may be of the greatest consequence to her health, her interest, and her happiness. – A physician who is a man of galantry, has many advantages in his endeavours to seduce his female patients; advantages but too obvious, but which it would be improper to recite. A physician who avails himself of these, is a mean and unworthy betrayer of his charge, or of [27] that weakness which it was his duty, as a man of honour, to conceal and protect.

Temperance and sobriety are virtues peculiarly required in a physician. In the course of an extensive practice, difficult cases frequently occur, which require the most vigorous exertion of memory and judgment, otherwise very important lives may often be lost. I have heard it said of some eminent physicians, that they prescribed as well when drunk as when sober. If there was any truth in this assertion, it contained the most severe reflection against their attention and capacity in their profession. It shewed evidently that they practised by rote, or prescribed for some of the more obvious symptoms, without attending to those nice discriminating circumstances, a knowledge of which makes the great difference between a physician who has genius, and one who has none. Intoxication implies a defect in the memory and judgment; it implies confusion of ideas, [28] perplexity and unsteadiness; and must therefore unfit a man for every business that requires the lively and vigorous use of his understanding. This state, however, is very different from that produced in some people by a chearful glass, which, when taken moderately, often enlivens and invigorates every active faculty of mind.

I may mention among the moral duties incumbent on a physician, that candor, which makes him open to conviction, and ready to acknowledge and rectify his mistakes. His obstinate adherence to an unsuccessful method of treating a disease, may sometimes be owing to a high degree of self-conceit, and a belief of the infallibility of a system. The error here is without remedy, because it flows from a weak and ignorant head. Profound knowledge and clear discernment may lead one into the extreme of diffidence and humility, but are inconsistent with self-con-[29]ceit. But it too often happens, that this obstinacy proceeds from a defect in the heart. A physician sometimes sees that he is wrong, but is too proud or rather too vain, to acknowledge his error, especially if the error is pointed out to him by another. To this species of pride, a pride incompatible with true dignity and elevation of mind, have the lives of thousand been sacrificed. A prudential regard indeed for the patient's safety may often make it necessary for a physician to conceal any embarrassment or mistakes from him, lest it alarm him and lose his confidence; but no man will persist in an error he is conscious of, who has the least regard to honour or principle.

III. I PROCEED now to make some observations on the decorums and attentions suitable to a physician, and such as tend most effectually to support the dignity of the profession.

Decorum, decency, and propriety, are words very indeterminate in their appli-[30]cation; for this reason, that the ideas annexed to them are partly founded in nature and common sense, partly in caprice, fashion, and the customs of particular nations. In the first case, the obligation to them is immutable, the same in all ages and nations; in the latter, it is fluctuating and less binding. When it is necessary I shall endeavour to mark this distinction.

I have already taken notice of the principal duties a physician owes to his patients, of the propriety of his attending to their tempers and constitutions, and allowing them every indulgence consistent with their safety. Sometimes a patient himself, sometimes one of his friends, will propose to the physician a remedy, which, for some reason or other, they expect may do him service. Their proposal may be a good one; it may suggest to the ablest physician, what, perhaps, till then, might not have occurred to him. It is un-[31]doubtedly, therefore, his duty to adopt it. Yet there are some of our faculty, who, from a pretended regard to the dignity of the profession, but in reality from the meanest and most selfish views, refuse to apply any remedy proposed in this manner, without regard to its propriety. But this is an insolent piece of tyranny. Every man has a title to speak where his life or his health is concerned, and every man is entitled to suggest what he thinks may save the life of his friend. It becomes them to make their proposals with politeness, and a suitable deference to the superior judgment of the physician; it becomes him to hear what they have to say with attention, and to examine it with candour. If he really approves, he should frankly say so, and act accordingly; if he disapproves it, he should declare his disapprobation in such a manner, as shews it proceeds from conviction, and not from pique or obstinacy. If a patient is determined to try an improper or dangerous [32] remedy, a physician should refuse his sanction, but he has no title to complain of his advice not being followed, as he has no right to hinder any man from going out of the world in his own way.

A physician is often at a loss in speaking to his patients of their real situation in respect to hazard to their lives. A deviation from truth is sometimes in this case both justifiable and necessary. It often happens that a sick person is dangerously ill, who, if he was to be told of his danger, would be hurried to his death. It sometimes again happens, that a man is seized with a dangerous illness, who has made no settlement of his affairs, and yet perhaps the future subsistence or happiness of his family may depend on his making such a settlement. In this and other similar cases, it may be proper for a physician, in the most prudent and gentle manner, to give warn his patient of his real danger, [33] and even solicit him to execute the necessary settlements. But, in all cases whatever, it is a physician's duty never to conceal his real situation from the relations. Justice obliges this, because it gives them an opportunity of calling for further assistance, if they should think it either necessary or decent. To a man of a compassionate and feeling heart, this is one of the most disagreeable duties in the profession: but it is indispensible. The manner of doing it requires equal prudence and humanity. What should reconcile him most easily to this painful office, is the reflection that, if the patient should recover, it will prove a joyful disappointment to his friends; and, if he die, it makes the shock fall more gently. Let me here exhort you against a barbarous custom of some physicians, the leaving your patients when their life is absolutely despaired of, and when it is no longer decent to take fees. It is as much the business of a physician to alleviate [34] pain, and to smooth the avenues of death, as to cure diseases. Even in cases where his skill as a physician can be of no further avail, his presence and assistance as a man and as a friend may be grateful and useful, both to the patient and his nearest relations. Neither is there any propriety in his going out at one door when the clergyman enters at the other; a quaint conceit of some of our faculty, more expressive of impiety than humour. On the contrary, it is decent and proper that they should mutually understand and assist one another. The conversation of a clergyman of chearful piety and good sense, in whom a sick man confides, may sometimes be of more consequence in composing the anguish of his mind, and the flutter of his spirits, than all the drugs in the dispensatory; while a gloomy and wrong-headed enthusiast may terrify him to distraction, and cut short a life, which, by proper attention, there was the greatest probability of saving. [35]

Every thing relative to the fees of physicians must be regulated by the customs of particular countries. That there should be a distinction, however, between the fees to be taken from a man of fortune, and from one whose circumstances are more narrow, is a very plain rule, founded in humanity and common sense; yet there is often great difficulty in its application. A depressed situation in life does not imply a mean nor a sordid spirit; neither does an opulent and elevated station imply a generous one. A noble mind with a small fortune, has generally too large a share or pride and sensibility, which are easily wounded by obligations not conferred in a very delicate manner. A Physician who does not take fees, should be particularly careful not to fail in point of attention to his patient; otherwise, instead of a favour, he does him an essential injury; an injury the more cruel, because the patient does not feel himself at liberty to complain of the neglect, or to call another physician. [36]

There are often unhappy jealousies and animosities among physicians, from which their patients may suffer severely. A physician, however, who has any sense of justice or humanity, will never involve his patient in the consequences of private quarrels, with which he has nothing to do. Physicians in consultation, whatever may be their private resentments or opinions of one another, should divest themselves of all partialities, and think of nothing but what will most effectually contribute to the relief of those under their care. If a physician cannot lay his hand to his heart, and say that his mind is perfectly open to conviction, from whatever quarter it shall come, in common honesty he should decline the consultation. Many advantages arise from two physicians consulting together, who are men of candour, and mutually confident of each other's honour. A remedy may occur to one which did not to another; and a physician may want resolution, or sufficient confidence in his own opinion, [37] to prescribe a powerful but dangerous remedy, on which, however, the life of his patient may depend; in this case the concurrent opinion of his brethren may determine his own conduct. But if there is no mutual confidence; if opinions are regarded, not according to their intrinsic merit, but according to the person from whom they proceed; or if there be reason to believe, that proposals delivered with openness are to be whispered abroad, and misrepresented to the publick, without regard to the obligations of honour and secrecy; and if, in consequence of this, a physician is singly to be made responsible for the effects of his advice; in such cases, consultations of physicians tend rather to the detriment than to the advantage of the sick, and end commonly in some very trifling and insignificant prescriptions. – The quarrels of physicians, when they end in appeals to the public, generally hurt the contending parties; but, what is of more consequence, they discredit the profession, and expose [38] the faculty itself to ridicule and contempt. – Nothing, in my opinion, but the cause I hinted above can justify any physician from refusing to consult with another, when he is required to do so. If he is conscious he cannot behave with temper, and that his passions are so rankled as to impair his judgment, he may and ought to refuse it. But such circumstances, as the place where the person he is to consult with had his degree, or, indeed whether he had a degree from any place or not, cannot justify his refusal. It is a physician's duty to do every thing in his power, that is not morally criminal, to save the life of his patient, and to search for remedies from every source, and from every hand, however mean, and in many respects contemptible. This, it may be said, is sacrificing the dignity and interests of the faculty. But I am not here speaking of the private police of a corporation, or the little arts of craft. I am speaking of the duties of a liberal profession, whose ob-[39]ject is the life and health of the human species, to be exercised by gentlemen of honour and ingenuous manners. The dignity of such a profession can never be supported by means that are inconsistent with its ultimate object, and that tend only to swell the pride and fill the pockets of a few individuals.

There have arisen from time to time, and particularly in France about twenty years ago, great disputes about the separate boundary of physic and surgery, and the proper subordination of surgery to physic. A dispute pernicious to the interests of mankind, and very unworthy of scholars and gentlemen. I shall take this opportunity of giving my sentiments concerning it.

There was anciently, as Celsus informs us, a division of medicine into three parts; the first regarded the regulation of diet; the second, the prescription of remedies; the third, manual operations. [40] The two first, though distinguished in theory, were always united in practice; the last has often been practiced separately. Sometimes the ancient physicians performed the manual part themselves, at other times, it was done by slaves kept in the house for that purpose. Among the moderns the arts of physic and surgery have often been promiscuously practised by the same persons; for example, Hildanus[14], Severinus[15], Bartholine[16], and many others of distinguished genius and literature. But in many parts of Europe, both formerly and at this day, surgery has not been reckoned among the liberal professions, but surgeons have ignominiously been classed with the corporation of barbers. The separation of physic from surgery in modern times, has been productive of the worst consequences. The physicians and surgeons, formed into separate societies, had separate interests to support, which, in many cases, clashed with each other. The surgeons claimed to themselves not only the exclusive pri-[41]vilege of performing all operations, but likewise the management of most external diseases, and some internal ones, where operations were supposed to be often necessary: by which means the method of cure in many diseases was directed by ignorant and illiterate people. But it must be apparent to every sensible and ingenuous observer, that the diseases of the human body are so intimately combined, that it is impossible to understand some of them perfectly and be entirely ignorant of all the rest; and equally impossible to understand any of them, without some knowledge of Anatomy, and the Animal œconomy, both in its sound and morbid state. It must at the same time be owned, that an able practitioner, well grounded in such general knowledge, may have considerable advantages, and more readily make improvements, by attaching himself to the study of one or two particular diseases.– Every disease, external as well as internal, falls under the cognizance of the [42] physician, and it is a reflection on him to be ignorant of any of them; neither is it possible to fix any such precise boundaries between external and internal diseases, as to render the distinction in any degree useful, or applicable in practice. Suppose a person to break his leg, and a fever and gangrene to ensure; the question occurs, whether the limb should be immediately amputated, or if the effects of certain medicines, given with a view to stop the progress of the gangrene, should be waited for till another day. It is evidently the business of a physician, in this case, to judge from the symptoms, from the habit of body of the patient, and from other circumstances of moment, whether the delay is prudent or not. – As to the performance of the operation itself, that is a different question. The genius, the discernment and education requisite to make a good physician, are not necessary to make a good operator. – What is peculiarly necessary to make a good operator, is a resolute, collected mind, a good eye, and a steady [43] hand. These talents may be united with those of an able physician; but they may also be separated. – If surgery were confined to a set of men who were to be merely operators, it might justly be expected, that the art would be more quickly brought to perfection by such men, than by those who follow a more complicated business, and practise promiscuously all the branches of medicine. The same advantage would accrue to pharmacy, if apothecaries were to be confined to the mere business of compounding medicines. But, in fact, this is not the case. In some parts of Europe surgeons act as physicians in ordinary, in others, the apothecaries do this duty. The consequence is, that in many places physic is practised by low, illiterate men, who are a scandal to the profession. On the other hand, whilst all the branches of medicine are indifferently practiced by men formed into separate societies, differently educated, and having different interests, it is plain that [44] none of the branches can be cultivated to the greatest advantage, and that the interest of mankind must often suffer from the jealousies and jarrings of professions, whose boundaries are not fixed, though they will much oftener suffer by the vilest collusion between the several professions. It is a known fact, that in many parts of Europe, physicians who have the best parts, and best education, must yet depend for their success in life upon apothecaries, who have no pretensions either to the one of the other; and that this obligation is too often repaid by what every one who is concerned for the honour of medicine must reflect on with pain and indignation.

From what I have said, I think it will appear, that I have no intention to throw reflections upon any particular profession. The profession of every branch of medicine is respectable, when it is exercised with capacity and honour. I only contend for an evident truth, either that [45] the different branches should be separately professed, or if one person will profess them all, he should be regularly educated to, and thoroughly master of them all. I am not here settling points of precedence or heraldry, or insinuating the deference due to degrees in medicine. As a doctor's degree can never confer sense, that title alone can never command regard; neither should the want of it deprive any man of the esteem and deference due to real merit. If a surgeon or apothecary has got the education and knowledge required in a physician, he is a physician to all intents and purposes, whether he has is a doctor or not, and ought to be respected and treated accordingly. In Great Britain surgery is a genteel and honourable profession. In most parts of it surgeons are the physicians in ordinary to most families, which their education and knowledge often gives them an undoubted title to be, and a physician is only called where a case is difficult, or attended with danger. There are certain [46] limits, however, between the two professions, certain forms of good breeding to be observed, which the gentlemen of both sides ought to attend to; as they are established by the customs of the country, and by the laws of their particular societies. But, I imagine, a physician of a candid and liberal spirit, will never take advantage of what a nominal distinction, and certain, real, or supposed privileges, give him over gentlemen, who, in point of real merit, are his equals; and that he will feel no superiority, but what arises from superior learning, superior abilities, and more liberal manners. He will despise those distinctions founded in vanity, self-interest, or the caprice of the world; and will take care, that the interests of science and of mankind shall never be hurt by a punctilious adherence to formalities.

Among the peculiar decorums of a physician's character, much regard has been laid on a certain formality in dress, [47] and a particular gravity and stateliness in the general course of his behaviour. I formerly observed, that decorum and propriety have their foundation sometimes in nature and common sense, sometimes only in caprice and fashion. This observation may be exemplified by the present subject. In many cases a particular formality and pomp of dress is highly proper, independent of any fashion whatever; for example, in judges and in magistrates. Whatever circumstances in their mode of dress, or external appearance, make them the objects of love and reverence, there undoubtedly are highly proper; because by a very natural association, they impress the minds of the people with a due veneration and fear of the laws. Neither is there any danger of abuse from this reverence procured to the office of a magistrate. The case is very different in the profession of medicine. There is no natural propriety in a physician's wearing one dress in preference to another; it not being ne-[48]cessary that any particular respect or authority should be annexed to his office, independent of what his personal merit commands. Experience, indeed, has clearly shewn, that all our external formalities have often been used as snares to impose on the weakness and credulity of mankind; that in general they have been most scrupulously adhered to by the most ignorant and impudent of the profession; that they frequently supplant real worth and genius; and that, so far from supporting the dignity of the profession, they often expose it to ridicule and contempt. If then there is no natural and real propriety in a physician's wearing a distinguishing dress, he can be under no obligation to use it, but what arises from the particular fashion of the country where he resides. This is an obligation, however, which common sense and prudence make it necessary he should regard. If the customs or prejudices of any country affix the idea of sense, knowledge, or dignity to any load of artificial [49] hair worn on the head, to a gold-headed cane dangling at the wrist, to a full-trimmed coat, and a sword, it is unquestionably a physician's business, from the common principles of self-preservation, to equip himself accordingly. But in a country where a physician's capacity is not measured by such standards, and where he may dress like other people, without sinking in their estimation, I think he is at full liberty to avail himself of this indulgence, if he so chuses, without being considered as deviating from the propriety and decency of his profession.

As to the general character of a physician's manners, I see no reason why they should be different from those of a gentleman. If the fashion of a country absolutely requires that he always look exceeding wise and solemn, he must submit; but if he voluntarily lays himself under these or any other restraints, except those of honour and conscience, and assumes a character which is not his [50] own, there is reason to suspect, that he either a knave or a fool.

There is great impropriety in a physician's indulging himself in a certain nicety and refined delicacy, which makes him liable to be disgusted with some disagreeable circumstances he must unavoidably meet with in his practice. Genuine delicacy is a virtue of the mind, and though it shews itself by an attachment to cleanliness, neatness, and even elegance, where it can be afforded, yet it always gives place and forgets itself, where duty or the interests of humanity require it. It is ridiculous in a physician to think any attentions, or any duties, below his dignity, which can contribute to the relief of his patient. When necessity requires it, he acts unworthily, if he do not become, to the best of his abilities, both surgeon, apothecary, and even nurse. If, however, without such necessity, he encroaches on another's province, then, indeed, he degrades himself; not because [51] he acts below the dignity of a physician, but because he behaves in a manner unbecoming the character of a gentleman.

Having freely expressed my sentiments concerning what I think a false dignity assumed by some of our faculty, I shall now, with the same freedom, animadvert on a circumstance not unfrequent in the behaviour of other learned men, physicians as well as others, which seems to me very essentially injurious to the true dignity of our profession; I mean that servility of manners and that abject flattery of people of rank and fortune, which, in modern times, so often disgraces men, in other respects eminent for learning and ingenuity. This is one of the most humbling circumstances in the general character of men of genius and society. – The external magnificence and splendor which surround high rank seems to dazzle the understandings of those who live at a distance from it, and who are ignorant of what a contemptible [52] inside it often covers. It makes them crouch to those outward distinctions of title and fortune, which their philosophy, if it was anything but a name, would make them despise, at the same time that they paid them that external respect, which the order of society, and the established forms of good breeding required. We should at first view expect to find men of high intellectual abilities, possessed of a very independent spirit, and great elevation of mind. Superior parts and knowledge would even seem to lay a natural foundation for pride, or a generous elation of heart, accompanied perhaps with too high a degree of self-esteem, upon a comparison with those who assume a tone of superiority upon the score of rank and fortune alone. If such a pride, however, is not attended with insolence, but is properly corrected by good nature or good manners, it is a very pardonable weakness, as it arises from qualities that ought to confer real rank and importance upon those who possess them. [53] But there is a pride of science, unaccompanied with true dignity, assuming and overbearing to inferiors, but meanly fawning and submissive to those of superior rank and fortune. – I will readily acknowledge, that there is a certain aukwardness unavoidably attendant on men of retirement and study; and that there is an ease, a gracefulness, and elegance of manners, which can only be acquired by a daily intercourse with the polite world. But the liberal manners, and liberal spirit, which still more peculiarly distinguish the gentleman, are confined to no rank or situation, as they have no dependence on external forms of good breeding; and should particularly be expected to prevail among those whose minds have been opened and enlarged in the pursuit of science.

Great disputes have arisen in our profession, about the propriety of a physician's keeping secrets or nostrums. It has been said, with great plausibility, in [54] vindication of this practice, that the bulk of mankind never attend, or pay any regard, to what is made level to their capacities; and that they put no value upon what costs them nothing. Experience certainly shews, that mankind are wonderfully attached to whatever has an air of mystery and concealment. A vender of a quack medicine does not tell more lies about its extraordinary virtues, than numbers of people do who have no interest in the matter; even men of undoubted sense and probity. A passion for what is new and marvellous, operates more or less on every human imagination; and, in proportion as that is heated, the understanding is duped and confounded. When a nostrum is once divulged and sold for a trifle, all its wonderful qualities immediately vanish, and in a few months it is generally forgot. If it is really a valuable medicine, the regular faculty may adopt it, but it never recovers its high reputation in the world. – It is likewise said, that this [55] is the only way in which any valuable medicine can be effectually introduced into practice; because it is the only way by which it can produce the attention of the publick, or a fair trial of its virtues; as the bulk of mankind will much more readily follow the directions of a man who professes to cure them by mysterious means, than a regular physician of established honour and capacity, who prescribes such plain things as their own common sense may shew them the propriety of. It is further alleged, that some of the best remedies in medicine were originally introduced as secrets, though discredited by the regular physicians. But allowing this to be true, yet I am persuaded, that these nostrums, on the whole, do much more hurt than good to mankind; that they hinder the advancement of the art, by making people neglect what is known and established, in pursuit of what is unknown and never to be divulged; that, by heating the imagination, they impose on the judgment, [56] and confound the science with a multitude of false facts; and that, from their being generally kept in the hands of worthless and illiterate men, who prescribe them indiscriminately, they are one of the greatest public nusances in Great Britain.– In some places on the continent of Europe, however, physicians of established honour and reputation keep nostrums. In such hands, the same abuses will not be committed, as we experience here; but still the practice has an interested and illiberal appearance.

I shall conclude this subject with some observations on a charge of a heinous nature, which has been often urged against our profession; I mean that of infidelity and contempt of religion. I think the charge absolutely false, and will venture to assert, that the most eminent of our faculty have been distinguished for regard to religion. I shall only mention, as examples, Harvey, Sydenham, Arbuthnot[17], Boerhaave[18], Stahl, and Hoff-[57]man[19]. – It is easy, however, to see whence this calumny has arisen. Men whose minds have been enlarged by extensive knowledge, who have been accustomed to think, and to reason upon all subjects with a liberal and generous freedom, are not apt to become bigots to any particular sect or system whatever. They can be steady to their own principles without thinking ill of those who differ from them; but they are particularly impatient of the authority and controul of men, who would lord it over their consciences, and to dictate to them what they are to believe in every article where religion is concerned. This freedom of spirit, this moderation and charity for those of different sentiments, have frequently been ascribed, by narrow-minded people, to secret infidelity, scepticism, or, at least, to lukewarmness in religion; while, at the same time, some men, who were sincere and devout Christians, exasperated by such reproaches, have expressed themselves sometimes in an unguarded man-[58]ner, and thus given their enemies an apparent ground of clamour against them. This, I imagine, has been the real source of that charge of infidelity so often and so unjustly brought against physicians. In a neighboring nation, where few people have been used to think or to reason with freedom on religion, and where, till of late, no man durst express himself with freedom on the subject, some ingenious and spirited writers have, within these few years, shone forth, who, impatient to shew their newly-acquired liberty, have attempted to shake the foundations of all religion, natural as well as revealed. Lately emancipated from superstition, by a transition not unusual, they have plunged at once into Atheism. It is happy for Mankind, that these people have carried matters this length; because evil must very quickly cure itself. Mankind may have their religious opinions diversified by various superstitions; but religion is natural to the human mind, and every attempt to eradicate it, is equally [59] wicked and impotent. But, supposing that Atheism came universally to prevail, together with the disbelief of a future state of existence of the immortality of the soul, and what has generally been thought to be intimately connected with it, of its immateriality, the duration of such sentiments would necessarily be very short; because they would at once unhinge all the bonds of society, and produce a continued scene of universal anarchy, wickedness, and despair. Yet I am sorry to say, that at present they are making a very alarming progress. Divested of that uncouth, metaphysical dress, under which they long lay concealed, the gloomy entertainment of a few recluse men, void of sensibility and abstracted from the business of humanity, they are now produced to the world, adorned with all the arts of eloquence, wit and humour, and perfectly adapted to the capacities of petit-maitres and chamber-maids. So far as they contain any argument, their futility has been often demonstrated a thousand times over; but [60] indirect hints, insinuations and ribaldry are unanswerable. The method taken by the present patrons of infidelity to propagate their opinions is extremely dangerous. With a matchless effrontery, they insinuate, that all who avow their belief in natural or revealed religion, are either hypocrites or fools. This is attacking youth on a very weak side. A young man, of a high and liberal spirit, disdains the idea of hypocrisy; and, from an ill-judged pride, is afraid of whatever may subject him to so mean an imputation. Vanity, again, is the most universally ruling passion among mankind, especially among young people, who commonly dread contempt above every thing, and resent any reflection on the weakness and narrowness of their understanding, more than any imputation on their principles or morals. But I will venture to affirm, that men of the most enlarged, clear, and solid understandings, who have acted in life with the greatest spirit, dignity, and propri-[61]ety, and who have been regarded as the most useful and amiable members of society, have never been the men who have openly insulted, or insidiously attempted to ridicule the principles of religion; but, on the contrary, have been its best and warmest friends.– Medicine, of all professions, should be the least suspected of leading to impiety. An intimate acquaintance with the works of nature elevates the mind to the most sublime conceptions of the Supreme Being, and at the same time dilates the heart with the most pleasing prospects of Providence. The difficulties that must necessarily attend all deep enquiries, into a subject so disproportionate to the human faculties, should not be suspected to surprize a physician, who, in his daily practice, is often involved in perplexity and darkness, even in subjects exposed to the examination of his senses. Yet such is the inconsistency sometimes found in characters, that we find examples of men disputing the evidence of the most interesting principles of [62] religion, who, in the business of common life, betray a childish credulity; and who embrace, with the most enthusiastic attachment, such theories, as are the mere sportings and vagaries of a lively imagination. – But there are some peculiar circumstances in the profession of a physician, which should naturally dispose him to look beyond the present scene of things, and engage his heart on the side of religion. He has many opportunities of seeing people, once the gay and the happy, sunk in deep retired distress; sometimes devoted to a certain, but painful and lingering death; sometimes struggling with bodily anguish, or the still fiercer tortures of a distracted mind. Such afflictive scenes, one should suppose, might soften any heart, not dead to every feeling of humanity, and make it reverence that religion which alone can support the soul in the most complicated distresses; that religion, which teaches to enjoy life with chearfulness, and to resign it with dignity. A physician, who [63] has the misfortune to be cut off from the happy prospects of futurity, if he has common good nature, will conceal his sentiments from those under his charge, with as much care as he would preserve them from the infection of a mortal disease. Fortified with insensibility, or ardent in the pursuits of business or pleasure, he may not feel in how forlorn and melancholy a situation he himself is placed, but it is barbarous to deprive expiring nature of its last support, and to blast the only surviving comfort of those who have taken a last farewell of every sublunary pleasure and connection. If motives of humanity, and a regard to the peace and happiness of society, cannot restrain a physician from expressing sentiments destructive of religion or morals, it is vain to plead the obligations of politeness, and the decency of the profession. The most favourable construction we can put on such conduct, is to suppose, that it proceeds from an uncontroulable levity of mind, or an [64] unbounded vanity, that forgets all the ties of morals, decency, and good manners.

I shall make no apology for seeming to go a little out of my way in treating of so serious a subject; because I think I stand in no need of one. In an enquiry into the office and duties of a physician, I thought it necessary to attempt to wipe off a reflection, which appeared to me derogatory to our profession; and, at the same time, to caution you against that thoughtless levity, or ridiculous vanity, in conversation, which may give ground to imputations of dissoluteness of principle, equally dangerous to society, and to your own truest interest and honour.

IV. I PROCEED NOW to explain the connections of the several branches of physic with the practical part of it, and to enquire how far a previous knowledge of these is necessary, in order to practise with reputation and success. [65]

Here I must previously observe, that though the whole works of Nature are so intimately connected, that no one part of them can be well understood by considering and studying it separately by itself; yet these links of the chain that are nearest to it, must be particularly attended to. In order to be qualified for the practice of physic, a variety of branches of knowledge, seemingly little connected, are extremely necessary. As this is the case, it is proper that a student should be on his guard not to waste his time and labour in pursuits which either have no tendency, or a very remote one, to throw light on the main ends of his profession. Human life is too short to leave room for every study that may be deemed ornamental to a physician; it will not even admit time for every study that has a remote connection with physic. Every one of the sciences I am going to name, considered separately are of infinite extent; but it will be necessary for [66] a physician to limit his application and attention to each of them, within certain bounds: he must confine his views to such parts of them, as are really subservient to the ultimate end of his profession. If a student's genius leads him to a particular attachment to any of these preliminary sciences, he may, if he pleases, indulge himself in the study of it, in its utmost extent and application; but then he should not impose on himself, and consider this as studying physic. – The study of physic is the study which qualifies a man for being a physician.

The absolute necessity of a previous knowledge of anatomy to the practice of physic, is apparent at first view, and needs no illustration.

The necessity of the knowledge of physiology, which comprehends the doctrine of the animal fluids, and of all the functions in their sound state, is equally evident. – When you inquire into [67] this subject, you find the human body a machine, constructed upon the most exact mechanical principles. In order, then, to understand its movements, you must be well acquainted with the principles of mechanics. Considering the human body in another view, you find fluids of different kinds circulating through tubes of various diameters; and therefore find, the laws of their motions cannot be understood, without a previous knowledge of the principles of hydraulics. In the same way, the eye appears to be a most admirable optical machine; and, the phenomena of vision are found inexplicable, without a knowledge of the principles of optics. – As the human body is surrounded with a heavy elastic fluid, the air, subject to various changes, in respect of gravity, heat, moisture, and other qualities which greatly influence the human constitution, it is proper to be acquainted with the nature and properties of this fluid; which makes the science of pneumatics. It were easy to [68] adduce many more examples, to shew how absolutely necessary a knowledge of the various branches of natural philosophy is to the right understanding of the animal oeconomy, both in its sound and morbid state.

But the different phenomena of the animal oeconomy are not all to be explained from common mechanical principles; various changes are induced upon the fluids, in consequence of chemical principles. It is, therefore, necessary to be acquainted with the chemical history of the animal fluids, with the chemical history of whatever is taken into the human body as food or physic, and, in general, of all the substances which can, in any degree, influence it. This shews the necessity of a knowledge of chemistry, previous to the study of the practice of physic.

Yet the most accurate knowledge of anatomy, and of the principles of me-[69]chanics and chemistry, are insufficient to explain all the phænomena in the body. The animal machine differs in many respects from an inanimate one. The former has a power of beginning motion within itself. An internal principle directs and influences the whole operations of the human body, by a set of laws totally distinct from, and independent of, any principles of mechanics or chemistry hitherto known. Many feeble and impotent attempts have been made to explain the phænomena of the animal body upon mechanical and chemical principles alone, but without the least shadow of success. The laws of the mental system are of the most difficult investigation; yet are equally steady and regular with any other laws of nature. An animal machine likewise differs from a common machine, in having a power, to a certain degree, of curing its own disorders, and of rectifying any deviations from its healthy state. As in the case of fractured bones, incarnation of [70] wounds, enlargement of one kidney when the other is destroyed, and in the successful efforts of nature in the cure of many diseases.

In order to illustrate the human physiology, a knowledge of the comparative anatomy of some animals, that most nearly resemble man, is extremely requisite. Several of the most important discoveries in the animal œconomy, have been made or illustrated by observations or experiments first made on brutes, many of which it was impossible to have made on the human subject, e.g. the experiments relating to the circulation of the blood, respiration, muscular motion, sensibility and irritability of different parts of the body, and the effects of various medicines. The instincts of brutes have often given the first hint of valuable remedies, and might throw light on the subject of regimen, and the cure of diseases, if they were properly attended to. At [71] the same time it must be acknowledged, that the comparative anatomy of other animals has often led into great mistakes, by too hastily applying it to the human body.

The writers on physiology have usually considered the body as a fixt, permanent subject, exhibiting uniformly the same appearances; but, in applying the knowledge of the animal œconomy to practice, it is necessary to consider the human constitution, as perpetually fluctuating, and not, perhaps, exactly the same in any two people upon the earth. It were endless to trace the infinite diversity of constitutions among mankind, neither would it be an enquiry of great utility; but there are some varieties which it is absolutely necessary to attend to. These depend chiefly on the difference of age, sex, climate, and manner of living; and some original temperaments, or habits of body, not produced by any of these circumstances. [72] It belongs to physiology, to enquire into the laws of the union between the mind and the body; into the effects of culture and education upon the constitution; into the power of habit, the effects of enthusiasm, and force of imagination. This short detail shews how extensive a study physiology is, and how intimately connected with the study of the practice of physic.

As physiology considers the whole appearances of the animal œconomy in its sound state, pathology considers the appearances in a morbid state. It delivers the general doctrine of the causes, effects, and symptoms of diseases. The therapeia treats of the general laws to be observed in the cure of diseases, and of the general nature of the remedies for that purpose. This includes surgery and the materia medica. The immediate usefulness of a knowledge of mechanics, appears most evidently in the practice of surgery. [73] This art has, in fact, received the greatest improvement within these hundred years, since the doctrine of mechanics came to be more generally understood.

A knowledge of the *materia medica* is intimately and immediately connected with the practice of physic. It contains the doctrines of the instruments with which a physician operates, and a history of the effects of medicines. In this branch the use of chemical knowledge is apparent. It teaches how to preserve and separate the useful parts of medicine. But in pharmacy, the knowledge of chemistry is indispensible. For want of this knowledge, at least for want of a proper application of it, true pharmacy has, till of late, been disgraced by the grossest blunders.

The effects of medicines on the human body are sometimes explicable upon mechanical, sometimes upon chemical, principles; but much oftener depend on the [74] effects they produce upon the nervous system. An enlarged knowledge, therefore, of mechanics, chemistry, and physiology appear necessary to a physician, in order to enable him to explain the phænomena of the animal œconomy, and likewise to explain the operations of remedies.

The science of botany is subservient to the practice of physic, so far as it facilitates the knowledge of plants, by reducing them into the most commodious and perfect system; and though it is not necessary for a physician to be particularly acquainted with the name and history of every plant he meets with, yet he ought to be so well founded in the principles of botany, as to be able to find its place in the system, and to describe it scientifically; and he ought to be acquainted with every material circumstance relating to those plants, which are either used in diet or as medicines. The same obser-[75]vation I have made, with respect to the knowledge of botany necessary to a physician, is equally applicable to every branch of natural history.

I have now shortly explained the connection of the several branches of physic, with the praxis medica, which comprehends the hygieine, or the method of preserving health and prolonging life, and the application of general pathology, and general therapeutics, to the history and cure of particular diseases. – It will naturally then be asked, is a person entirely unqualified for the practice of physic and surgery, who is not master of all these branches of learning, which have been alledged to be necessary preliminaries? To this it may be answered, that one may, in some measure, practice physic as he may do a mechanic art, without any knowledge of its principles. A sailor may navigate a ship, who is ignorant of the principles of navigation; and a person may construct a dial, who [76] knows nothing of the principles of astronomy, spherical trigonometry, or the projection of the sphere. It is the same in all the other practical arts of life; and yet in all these, there are obvious advantages arising from a knowledge of the principles on which they are founded. But in medicine, the necessity of being acquainted with the principles of the art is much greater; because there can be no general rules laid down for the practice of physic, which can be applied in all cases. Difference of age, constitution, climate and a thousand other circumstances, occasion necessary variations in the application of the most distinct rules that can be prescribed; and without a knowledge of the principles of his profession, and without extensive medical erudition, a physician must be at the greatest loss in making these deviations. It will be readily acknowledged, that there have been many physicians, eminently successful in practice, who, at the same time, were deficient in [77] the knowledge of the foundations of medicine. But this has been owing to their uncommon genius and sagacity, which enabled them to apply what little knowledge they had with judgment, and consequently with success; while, perhaps, another physician, of very extensive reading and knowledge, for want of this natural genius and sagacity, has blundered egregiously in his practice, by a wrong application of his knowledge, or by not knowing how to apply it at all. Besides, as medicine is so complicated a science, many of those who study it regularly, take a particular attachment to some of its preliminary branches, and these so far engage their attention, that they neglect their application to medicine, and likewise the study of the other branches. In consequence of this, some of our profession have been distinguished anatomists, chemists, and botanists, who, notwithstanding, have been very indifferent physicians. But surely it cannot thence be inferred, that their ill [78] success in practice was owing to their skill in these sciences, which must be acknowledged by every man of sense and candor, to be highly useful to them in their profession as physicians.

I do not insist here on the absolute necessity of a minute knowledge of these sciences, in their utmost extent; nor indeed could time be spared to acquire it. A particular acquaintance with the swell and appearance of the muscles, in all the various motions and attitudes of the body, is a study more necessary to a painter, or to a statuary, than to a physician; and, in this view, they ought to be the greatest of all anatomists. If chemistry is prosecuted, in its fullest extent and application, to all the useful and polite arts, it is a study that is boundless. So is botany, if one thinks it necessary to be acquainted with every circumstance relating to every plant indiscriminately that grows on the surface of the earth. It is therefore necessary, that a student, while [79] he endeavours to make himself master of the leading and fundamental principles of these sciences, should always have an eye to their particular application to his own profession, and bend his particular attention to that quarter.

On the whole, I hope it will appear sufficiently clear, that a physician, who understands the principles of his profession, who has an extensive acquaintance with every branch of natural knowledge connected with it, who properly applies his knowledge, and who has equal natural genius, and equal attention to practice, must have an infinite advantage, as a practical physician, over one who is ignorant of the principles of medicine, and of every science connected with it. Genius and sense are, indeed, the peculiar gifts of Heaven, and cannot be acquired by the most extensive learning, or the greatest efforts of industry. But, with these assistances, genius and sense are capable of the highest improve-[80]ments; and without them, the finest parts will turn to little account, either to the public or the individual.

Besides the above-mentioned branches of learning, which are in a manner essential to qualify one for the rational practice of physic, there are others, which, though perhaps rather ornamental, a physician, who aims at having a liberal education, should not be ignorant of. I hope I have no need to enforce to you a thorough acquaintance with the Latin language. A physician's reading must be confined within very narrow bounds, who is unacquainted with what has been the universal language of the learned in Europe for so many ages, and which serves to communicate their sentiments, from one nation to another, so easily and so quickly. The interests of learning will very soon suffer by its disuse, and by the present fashion of authors writing in their own native language. But I must here take notice of an error, [81] which they who value themselves on their knowledge of Latin are apt to fall into; and which has contributed, beyond any thing, to this growing evil. What I mean, is, too great anxiety about classical purity, and elegance of expression. The intention of language is to convey our ideas with clearness, force, and precision. If these can be joined to a style truly classical, it is a great additional beauty; but, from the numerous improvements made by the moderns in all the arts and sciences, there have arisen many ideas and objects, which the Roman classics could have no expressions for; because they did not know them. An author, therefore, who has occasion to express these ideas, is under a necessity of latinizing words in his native language, in order to express his meaning, or of adopting Latin words used only by authors of inferior note. If he is determined to use no phrase but what is strictly classical, he must often suppress altogether what he wish to say; [82] or if, by aukward straining, he grasps at a meaning, what honour he gains by eloquence of diction, he must lose in the more substantial points of energy, precision, and perspicuity. We have the peculiar felicity in our profession, of having a standard in Celsus, for purity and elegance of medical Latin; but there are still a variety of medical ideas to be expressed, which neither Celsus, nor any Roman writer ever dreamt of.

I must here recommend to you the study of that copious, expressive, and harmonious language, the Greek. Our oldest, and some of our best authors wrote in Greek; particularly Hippocrates, the father and founder of medicine. Almost all the medical terms of art are Grecian: a knowledge, therefore, of that language must evidently facilitate your progress in your own profession. Besides, it is not very decent for a physician, liberally educated, to be in the [83] daily use of terms to whose original he is a stranger.

The necessity of a knowledge of the French is apparent. Almost all the French authors, many of whom are very valuable, write in their own language; it is likewise become so universal in Europe, that every gentleman who travels must necessarily make himself master of it.

It may appear at first view superfluous, to recommend an attention to your own language. But it is notorious, that many physicians of real merit, have exposed themselves to the ridicule of the world, by their ignorance or inattention in respect to composition. It might be expected, that every one who has had the education of a gentleman, should write his native language, with at least grammatical exactness; but, even in this respect, many [84] of our writers are shamefully deficient. Elegance is difficult to attain; and, without great taste, dangerous to attempt. What we principally require in medical writings, is the utmost degree of perspicuity, precision, simplicity, and method. A flowery and highly-ornamented language in these subjects, is entirely out of its place, and creates a very just suspicion, that an author is rather writing from his imagination, than copying from nature. We have many bulky volumes in medicine, which would be reduced to a very small compass, were they stripped of their useless prefaces, apologies, quotations, and other tawdry ornaments, and confined to the few facts they contain, and to close inductive reasoning.– What I would principally recommend to you in every species of medical writing, next to a simple and candid history of facts, is a strict attention to method. I am no admirer of that pedantic display of system and arrangement, so remarkable in some of the Ger-[85]man writers, who split their subject into endless divisions and subdivisions. This may strike a reader, not accustomed to such kind of writing, with an high opinion of the author's ingenuity and accuracy; but in general it is a mere deceit. It is a mode of writing easily attained, and was in the highest perfection when the scholastic logic, which consisted rather of nominal than real distinctions, was held in general admiration. Yet this ostentation of method, even when carried to the greatest extreme, is highly preferable to the present fashionable way of writing in Great Britain, which seems to set all order at defiance. In the one style of writing, what is important, what is defective, and what is erroneous, is more easily detected; in the other, there is such a profusion of words, such a promiscuous jumble of facts and reasoning, and wit and flowers of rhetoric, that it requires a very attentive perusal of a book from beginning to end, to find out whether it is worth perusing or not. [86]

It would require too much time to enumerate all the other qualifications that might be deemed ornamental to a physician. In general, there is no reason why he should be excluded from any amusement, or any genteel accomplishment, that becomes a gentleman. On the contrary, these give an agreeable relaxation from the severer studies and fatigues of his profession; they render his conversation more chearful and entertaining; and, instead of that aukward pedantry, which modern men of learning have generally chosen to distinguish themselves, they diffuse a liberal, ingenuous, and liberal air over his whole manner. [87] [88 blank]

LECTURE II

The works of Nature are of infinite extent and variety: but, amidst all this variety, there is, such an intimate connection, that no one part can be thoroughly understood by studying it entirely detached from the rest. In our enquiries into the various branches of the works of nature, there are certain general views, and certain general principles of investigation, to be particularly attended to. The general views to be attended to, in the study of Nature, respect, 1. The advantages it brings to individuals. 2.

Public utility.

1. The advantages to individuals that attend enquiries into Nature, are suffi-[89]ciently obvious. They give exercise to many of the active powers of the mind; they gratify curiosity, the love of truth, and of whatever is great, beautiful, or wonderful: principles deeply implanted in human nature.

2. In regard to public utility, they promote all the useful and elegant arts, all the arts that tend to the happiness and ornament of human life. A profound knowledge of nature extinguishes pride and self-conceit, by rendering men more deeply sensible of their ignorance, their errors, and the very limited state of their faculties. It is favourable to the interest of religion, by exhibiting the most striking proofs of the infinite wisdom, power, and benignity of the Supreme Being, who supports this wonderful frame of things, by laws, often, indeed, unsearchable in their nature by human wisdom, but steady and uniform in their operation, and admirably fitted to promote the happiness of his creatures. Such a know-[90]ledge must impress every heart, endued with the least portion of sensibility, or not strangely perverted, with that awful veneration, that love and gratitude to the Divinity, that submission to his providence, and that reliance on his goodness, which alone constitute the soul of devotion. It has been imagined by some, that very extensive knowledge leads to Atheism; but there is not the least reason for such a suspicion. A little learning is, indeed, a dangerous thing to a weak and conceited man, who, from a superficial acquaintance with second causes, is apt to overlook the first and great cause. But to a sound understanding, extensive knowledge is the truest teacher of humility; it shews how often men are deceived in their supposed acquaintance with second causes; and that, even where many of these are clearly ascertained, yet, in tracing the chain that connects them, the most acute and profound genius must stop somewhere, and at last refer them to a supreme intelligent cause. [91] While we attempt, however, to clear philosophy from the charge of impiety, a very important distinction must be attended to. I will venture to maintain, that those philosophers have been the firmest supporters of religion, who have employed their genius and application in the investigation of the works of nature, and whose views in science have been grand and extensive. Among a multitude of examples I could bring to prove this assertion, I shall mention only three of our own countrymen, Lord Bacon, Mr. Boyle, and Sir Isaac Newton. Those philosophers, on the other hand, who have been the most distinguished propagators of Atheism, have been men not

much acquainted with the works of Nature, who searched for truth in their own little minds, not in the great world without them; men who, in regard to science and the useful arts, have either neglected them altogether, instead of promoting them by observation and experiments, or corrupted them by metaphysical sub-[92]tleties, often indeed ingenious and plausible, but that lead to no useful discoveries or improvements.

II. There is no branch of natural knowledge so useful or interesting, as that which relates to the human species; which is evident, when we consider that it includes,

1. Medicine, or the art of preserving health, of prolonging life, or of curing diseases.

2. The arts of improving the different faculties of the human body; as strength, agility, and endurance of pain, cold, hunger, and the many other evils mankind are subjected to.

3. The preservation and improvement of beauty.

4. The laws of union between the mind and body, and the mutual influence they have upon one another. This is [93] one of the most important enquiries that ever engaged the attention of mankind, and almost equally necessary in the sciences of morals and of medicine. It comprehends,

(a) The doctrine of the preservation and improvement of the different senses, external and internal, the memory, imagination, affections, and judgment.

(b) The history of the power and influence of the imagination, not only upon the mind and body of the imaginant, but upon those of other people.

(c) The history of the several species of enthusiasm.

(d) The history of the various circumstances in parents, that have an influence on conception, and the constitution and characters of their children.

(e) The history of dreams, with a view to our acquiring a power over them. [94]

(f) The history of the power and laws of custom and habit.

(g) The history of the effects of music, and of such other things as operate upon the mind and body, in consequence of impressions made on the senses.

(h) The history of natural signs and language, comprehending the doctrine of physiognomy and outward gesture.

I mention these only as a specimen, and not as a full enumeration of the many important articles contained under the natural history of the human species. I mention them as examples of the general views to be regarded in our investigation of nature, and very essentially connected with the science of medicine; but have taken no notice of the enquiries that relate to man in his moral, political, or religious capacity, as being foreign to my profession. [95]

III. I proceed now to lay down certain general principles, which require our attention in the investigation of nature, and shall endeavour to apply them more particularly to the science of medicine. When we look around us in the world, we find objects connected together, in a certain invariable order, and succeeding one another in a regular train. It is by observation and experience alone, we come to discover this established order and regular succession in the works of nature. We have all the evidence that the case admits of, to persuade us that nothing happens by chance: on the contrary, we have all possible reason to believe, that very event happens in consequence of an established and invariable law; and that, in cases perfectly similar, the same events will uniformly take place.

IV. Here I must observe, that, antecedent to all reasoning and experience, there is an original principle implanted in the human mind, whereby it is led to [96] a belief of the regular course of nature. In consequence of this principle, whenever a child sees any event succeeding another, he has an instinctive persuasion, that the same event will succeed it afterwards in the same circumstances. This persuasion does not flow from any connection he sees between the causes and the effects, nor from experience, nor from reasoning of any kind. So ardently do we desire to find every thing that happens within our observation, thus connected with something else, as its cause or occasion, that we are apt to imagine connections upon the slightest grounds: and this weakness is most remarkable among the ignorant, who know least of the real connexions established in Nature. – A principle of credulity seems likewise to be an original instinctive principle of the human mind, by which we are disposed to believe, prior to experience, not only the language of natural signs, but also the language of artificial signs, as soon as they come to be understood. Hence [97] credulity so peculiar to children, who at first believe every thing that is asserted to them to be true; and it is experience alone which teaches them to correct this original principle of belief. Dr. Reid has treated this subject with great acuteness, in his ingenious Enquiry into the human mind.

V. We obtain experience, either by the evidence of our own senses, or by the testimony of others.

1. The testimony of our senses, though generally considered as the highest degree of evidence, often deceives, and often fails us. The sensations excited in us, in consequence of impressions made on our organs of sense, depend,

(a) On the state of the medium through which the communication between the objects and the organs of sensation is made, e.g. the state of the air, when we judge of visible objects. [98]

(b) On the state of the organs of sensation themselves, every one of which may be vitiated in a variety of ways.

(c) Our unassisted senses often fail us, on account of the subtlety or minuteness of bodies, too quick or too slow motion, the object being too common, and many other causes.

(d) Although the impression is properly made on organs that are in their sound state, yet the ideas conveyed thence to the mind, may be so varied and modified by the imagination, as to mislead the judgment entirely. Thus every part of natural history, and medicine above all others, is overwhelmed with facts, attested by eye-witnesses of undoubted honour, which, notwithstanding, had never any existence but in their overheated imaginations.

2. The experience which we trust to from the testimony of others, is liable to the same imperfections with our own [99] personal experience, and often to the additional inconvenience, of our uncertainty of the accuracy or honesty of our authors.

VI. Having examined the sources of experience, I shall now proceed to consider the manner in which mankind agree in applying it. I have already attempted to explain the principles that lead men to believe that, what they have seen to happen in one case, will happen again in the same circumstances, and that the same causes will always produce the same effects. Whatever are the principles of their belief, the fact is true, both with regard to the most profound philosopher, and the most ignorant peasant. The only difference between these two consists in this; that the peasant concludes two cases to be precisely alike, because they resemble one another in their most obvious appearances; the philosopher, on the other hand, from a more enlarged experience and observation, does not so easily [100] trust to obvious appearances; he is aware of the various sources of deception, and therefore examines every minute and latent circumstance, before he ventures to pronounce the same judgment; and the difficulty of ascertaining, with precision, the exact similarity of cases, makes every true philosopher extremely sceptical in forming conclusions of what will happen, from what he has seen happen. – An African, who has seen water in an infinite variety of circumstances, but still retaining its fluidity, concludes, that fluidity is essential to water, and looks on it as lie, when he is told, that in certain parts of the world, water often appears in a solid form. His mistake does not proceed from trusting to experience, but from thinking he had experience, when in reality he had none. All that he could justly infer from his experience was, that water, in all the circumstances under which he had seen it, would remain fluid. But water, exposed to a degree of cold sufficient to congeal [101] it, was a circumstance in which he never saw it; therefore his experience could never tell him, what effect that degree of cold would have upon the water, whenever it came to be exposed to it. We have a remarkable instance of the effects of trusting to a partial and limited experience, in that firm belief which people ignorant of medicine so frequently have in the wonderful effects of particular remedies, especially if they are kept as secrets. Many an old woman, and, what is more surprising, many a grave philosopher, have infallible cures for a number of diseases, which every physician finds to be incurable. No physician indeed has the comfort of thinking himself possessed of an infallible cure, even for the scratch of a pin.

VII. Although facts afford the only solid foundation for genuine science, yet, when we consider them as unconnected with any other, they convey but little instruction. The phænomena of nature are infinite, but the capacities of [102] the human mind, and particularly the memory, is very limited. If these phænomena, therefore, were not reducible to certain general principles or laws, our experience of particular facts could do us but little service. But there is an instinctive strong propensity in the human mind, to be delighted with analogies, to compare and connect facts that resemble one another, and by this comparison, to reduce them to certain general rules, to apply such general rules to account for other effects, or to direct us in the production of them. The business of true philosophy is, in compliance with this natural propensity, to discover these connections, and to reduce them under certain general rules or principles, called laws of nature, by which we mean nothing else, but the most general facts relating to the operations of nature, which include a great many particular facts under them. The propensity of mind to reduce particular facts to general laws, appears from the anxiety which men shew to discover the cause of [103] any uncommon event. The discovery of this cause infers no more, than the discovery of that law of nature, by which the event is produced; for we mean nothing by natural causes, but certain general facts or laws, with which, what we call effects are uniformly and invariably connected as consequences: but the nature of this connection we are perfect strangers to. In our inquiries into nature, after we have arrived at the knowledge of some general laws, by an accurate comparison and arrangement of observations, we may, by comparing these laws together, discover laws of a more general nature; and thus, by a slow and cautious induction, we make advances to a knowledge of the most general laws, that regulate the system of nature, in all the different departments of the arts and sciences. But many obstacles concur to prevent the establishment of genuine philosophy upon this solid foundation; some of which I shall endeavour to explain. [104]

1. The anxiety and impatience of mankind to reduce all knowledge, and to refer all events to certain general laws, makes them unwilling to submit to this slow, but sure, method of investigation. They attempt, therefore, a shorter way of establishing those laws, in which they are misled, either by a loose reasoning from imaginary analogies, or by supposing the laws of nature to be fewer and simpler than they really are. The consequences of which are, the hasty reduction of the sciences into systems, imperfect and corrupted in all their parts.

2. The pleasure that men have in discovering analogies, makes them often fancy resemblances between things, where in truth there are none, or none of any consequence. Arguments from analogy very readily present themselves to a heated and fruitful imagination, while more direct and conclusive arguments, drawn from observation and experiments, often require painful attention and application, [105] and perhaps, after all, are insufficient to establish the wished-for principle or doctrine. I shall readily acknowledge the usefulness of analogies; they often facilitate the conception of things, which, without their assistance, could not easily be comprehended. It is likewise by reasoning from analogies, that we are most commonly led to the anticipations of the most useful principles and discoveries. But we ought never to acquiesce in analogies, when we can have access to more direct evidence; as all that analogies can lead to, are merely probable conjectures, commonly called theories, but more properly hypotheses.

3. There is a certain intoxication, that usually attends the supposed discovery of general principles in science, or useful inventions in the arts, which renders men of warm and lively imaginations altogether blind to every difficulty that lies in their way, and often induces them artfully to suppress them. The sup-[106]pression of facts that appear to contradict a

favourite hypotheses, is not always owing to want of candour in the author. Sometimes the author does not see them, sometimes he despises them, and sometimes he conceals them, from the fear of giving people an unreasonable prejudice against what he thinks an important discovery. Every true philosopher, however, will be particularly jealous of himself in this respect; and whenever he gets a view of a theory, will immediately set his invention at work, to contrive every possible experiment and mean of proof that can bring a direct and conclusive evidence, either of its truth or falsehood; and till such time as he can find such evidence, he considers his theory in no higher point of view than a probable conjecture.

This philosophical diffidence is so far from discouraging the investigation of causes and general laws, that, on the contrary, it greatly promotes it. A state [107] of suspense is always a disagreeable one, and the uneasiness it gives, becomes a powerful incitement to such further enquiries as may remove it. A zealous attachment to theories, may not only lead into dangerous mistakes, but by betraying men into a false security, cut off every motive to farther enquiry; representing it as an unnecessary piece of trouble. It is not philosophical scepticism, nor a humble opinion of our present knowledge, which checks the spirit of enquiry into the laws of nature; it is a mean opinion of the human powers, which effectually chills the ardor of genius, and blasts all grand and extensive views of improvement. In works addressed to the heart, that coldness and severe precision, so necessary in the investigation of truth, have no place; fancy there is in her proper element, and the loosest and wildest analogies may often be properly admitted. A philosopher may read a fairy tale with great delight, without reflection upon [108] his taste or understanding; but it reflects severely upon both, if he read with the same pleasure a philosophical investigation, not founded in observations and experiments, but in the vagaries of a lively imagination, unless he is sensible of its being a romance, and only allow himself to be charmed with the spirit or elegance of the composition.

4. There is a species of self-deceit upon this subject, which deserves particular notice. We often find those people inveighing bitterly against theories and hypotheses in philosophy, who are notoriously addicted to them, though not conscious of it themselves. This is remarkably the case with medical writers, who commonly abuse all reasoning and principles in physick which differ from their own, equally idle theory; and frequently declaim against theory in so vague a manner as would seem to condemn all reasoning and investigation of causes and principles as

useless and even perni-[109]cious. But it should be considered, that we cannot advance a step in the pursuit of knowledge, without reasoning. In every useful experiment, and especially in conducting a train of experiments, we must employ our reason; there must be some point in view, some anticipation of a principle to be established or rejected, and reason must determine all the circumstances to be attended to in making every observation, or experiment, with a view to ascertain this. Without reasoning, or without trusting to certain principles, either fully established, or rendered highly probable, we could never be benefited by experience, because we could never transfer it from the case we have seen, to the case immediately before us. For instance, I have a patient in an intermitting fever, which I propose to cure by the Peruvian Bark. I shall suppose I have cured five hundred patients by this medicine formerly; but yet I know I never cured one whose circumstances, in respect of age, temperament, and every other particular, exactly corresponded to [110] the one before me. If therefore I give the bark, I must reason, by tacitly adopting this principle, that the bark will universally cure agues, notwithstanding they differ in some circumstances. But this is a principle of which I have no direct and conclusive experience, but a principle which I have adopted, by a probable reasoning from analogy: and, indeed, it is not universally true, though physicians must proceed upon it in their practice, till such time as future observation shall ascertain the exceptions to it. Boerhaave, Hoffman, Stahl, and every systematic writer exclaim against theories, meaning one anothers theories; for each of them exclaim, though in different, and often opposite, manners, the proximate cause of every disease they give an account of, and the mode of operation of every remedy they prescribe, upon principles entirely hypothetical. Even Sydenham, though reckoned a purely practical writer, is full of hypothetical reasoning, which, however, had not the [111] usual effect of making him less attentive to observation; and, indeed, his hypotheses seem to have sit so loosely about him, that either they did not influence his practice at all, or he could very readily abandon them, and adopt new ones, whenever they would not bend to his experience.

VIII. It should seem, upon the whole, that all physicians must reason, and that the only difference among them consists in this, that some reason better than others. Some, for example, search into the causes of diseases, and the effects of remedies. Deeply sensible of the difficulty of the enquiry, and the various ways in which they may be deceived, they collect and arrange all the facts relating to the subject; when they have got a remote view of a leading principle, they attempt to bring a direct and conclusive proof, by experiment, of its existence. If the proof turn out against it, they see, and candidly acknowledge, that there is an [112] error somewhere; if the case does not admit of a direct proof, they consider their principle as only more or less probable, but never relinquish the pursuit. These, I think, have a just claim to the title of rational physicians. Others, upon the foundation of a few facts, and vague analogies, create a system of hypothetical principles; a creative imagination supplies materials, where they seem wanting; they employ all their ingenuity to twist facts into a correspondence with them, and such as will not bend to their purpose, they either suppress or reject, as incredible. In their practice, they neglect particular observation; as they consider their general principles so thoroughly established, as neither to stand in need to confirmation, nor to be capable of refutation. Such people dignify themselves, with the title of rational and dogmatic physicians. But surely every system-builder from the days of Hippocrates downwards, who has inlisted himself among [113] the rationalists, cannot have a right to this title; because many of their systems are different and contradictory. If one, therefore, is rational, all the rest must of course be contemptible. The truth is, nothing can be more absurd, than a physician's directing the method of cure in a disease, in the full confidence of his knowledge of the proximate cause, and the manner in which his remedies destroy the cause, when, in reality, he knows nothing of either. Some other physicians, who are stiled empirics, go upon a different plan: they lay it down as a principle, and a very false one it is, that the enquiry into the causes of diseases, especially proximate causes, and the enquiry into the manner in which remedies produce their effects, are both useless. Upon the experience of the success of a remedy in some particular cases, they venture to prescribe it indiscriminately in all others, where some of the most remarkable symptoms correspond, without any further enquiry into the circum-[114]stances in which they differ, or any other circumstance than can throw light on the nature of the disease. Here people deceive themselves, if they think they practice on the solid basis of experience. They are as much addicted to hypotheses, though of a different kind, as the dogmatists, and reason as absurdly. They are equally confident of principles, which are either utterly false, or true only in certain circumstances, and proceeding either from a luxuriant imagination, or a loose, erroneous induction from a few particular observations.

IX. I OBSERVED before, that in our enquiries into human nature, an impatience to acquire a knowledge of her laws, and a natural love of simplicity, make us to think them fewer and simpler than they really are. Enlarged knowledge certainly discovers that the laws of nature are perfectly uniform, and amazingly simple, if we compare them with the infinite extent and variety of her [115] works; but yet we must not think that they are confined within the narrow circle of our knowledge, or even our comprehension. When Sir Isaac Newton, towards the end of the last century, demonstrated, by a happy effort of genius, that all the planets gravitate towards the sun, by the same laws, and in consequence of the same principle, by which bodies on the earth gravitate towards its centre, many phænomena came to be explained by this simple law of gravity, of which formerly no account could be made. But it soon came to be applied to the explication of other phænomena, which afterwards appeared to happen in consequence of very different principles.* [*"Dr. Reid" at bottom p. 116] Des Cartes founded his system of the material world upon two principles, the existence of matter, and a certain quantity of motion originally impressed upon it. These two principles, however, were found insufficient, and it [116] has been made evident that, besides these, we must also admit the principle of gravitation just mentioned, cohesion, corpuscular attraction, magnetism, electricity, and other centripetal and centrifugal forces, by which the particles of matter attract and repel each other. Even Sir Isaac Newton was led by analogy, and the love of simplicity, to conjecture, but with singular modesty and caution peculiar to him, that all the phænomena of the material world depended upon attracting and repelling forces in the particles of matter. But we have reason now to believe, that in this conjecture he was deceived: for even in the unorganized kingdom, the powers by which salts, crystals, sparrs, and many other bodies, concrete into regular forms, can never be accounted for by attracting and repelling forces in the particles of matter; and in the vegetable and animal kingdoms, there are evident indications of powers, of a different nature from those of unorganized bodies. [117] We are conscious of an internal principle, which feels, which thinks, and which seems to be the original source of animal motions. We are, in a great measure, ignorant of its nature; but we know, that it has a system of laws peculiar to itself, and that, in consequence of its union with the body, certain effects are produced, which neither the laws of matter, nor of the mental system, considered separately, are able to explain.

X. We may here observe, how the different natural dispositions of men influence their literary character. We generally find men of lively imaginations, of keen and warm tempers, most disposed to attend to analogies, and the resemblances of things, in which fancy often deceives them. From these they are too ready to establish general principles, and to be so zealously attached to them, as not to see the objections that lie in their way. If, however, by any accident, their belief in the certainty of [118] their principles comes to be staggered, they instantly relinquish them, while, perhaps, they may be very well founded, only embarrassed with some difficulties, which a little more patience and temper might have conquered. To such people, the world often owes some useful discoveries, although they are seldom rewarded according to their merits. They are ruined by projects, sometimes because they are too extensive, and beyond their abilities to execute; sometimes because they have overlooked some small circumstance necessary to their successful execution, which a very dull man, afterwards observing, robs them both of the honour and profit of their inventions. In fine, this exhuberance of genius is generally attended with an impatience, that renders them incapable of a steady attention to observation and experiments, and a restlessness, which prevents their bringing any work to a conclusion and makes them desert it in the middle, [119] or after the most difficult part of it is conquered, in pursuit of some new plan.

There is a species of genius very much the reverse of the former, calm, sedate, discriminating, that attends very accurately to the differences of things, seemingly alike; that watches the excursions of more lively and inventive spirits, and too often exposes their mistakes to an undeserved ridicule. There are so few men of original genius who think for themselves, and strike out new paths in science, that they should meet with all possible encouragement, particularly when they propose their opinions with modesty, and becoming regard to those from whom they differ. Men who go often out of the common road, must sometimes go astray; but as they frequently make important discoveries, their errors ought to meet with a great deal of indulgence, even though for some time they should be warmly persisted in, [120] in consequence of that enthusiasm peculiar to this kind of genius. These two dispositions I have been speaking of, are often found united in the same person in different degrees. One may possess that warm and lively imagination, so peculiarly fitted for invention, and, at the same time, a clear, accurate, and collected judgment, that can distinguish with great precision; that discerns every objection to his proposed plans; and that, according to the real weight of evidence, can either reject them altogether, or preserve his mind in a proper degree of suspence. This union, where genius and sound understanding are so happily blended, and which seldom takes place, constitutes a philosopher of the first class and dignity.

But there is a numerous class of men, of considerable use in the learned world, who are mere drudges in science, who neither discern resemblances nor differences of things, who have no views [121] of principles, or plans or any sort, nor any ideas of method and arrangement. Some of these men are industrious and painful compliers; some attend to observations and experiments, with great patience and assiduity, though these, under their own conduct, are so trite, so vague, or so inaccurate, that they seldom lead to any conclusions, or can be safely trusted. Their labours, however, when under the direction of men of genius, may be rendered highly useful.

XI. In collecting a natural history, which is supposed to be subservient to the useful arts, and to be the foundation of a useful natural philosophy, it is necessary to make a selection of facts, among the infinite number with which the volume of nature presents us. Our views should then be confined to those, which, being compared and arranged, may lead to general and useful principles. The history, therefore, of any extraordinary production of nature, which has [122] nothing similar or analogous to it, is of little consequence but to gratify curiosity. Yet this principle of curiosity, and love of the marvellous, is so prevalent among mankind, that all the extraordinary events, and lusus naturæ, are what principally attract their regard. If a puppy come into the world with three or four heads, we have presently a very accurate description of the monster recorded in all the literary journals of Europe, though it is not a matter of the least consequence to mankind, whether the creature had four or forty heads. This love of the marvellous is very conspicuous in most writers of medical observations. We find them recording extraordinary cases, such as have nothing similar to them, such as never happened before, and, consequently, such as will probably never happen again, with a tiresome minuteness of description; while the symptoms that discriminate some common diseases from others of a different nature, which resemble them, [123] are far from being yet ascertained, although the lives of thousands have been lost from these distinctions not being established. I do not mean here to object to the recording extraordinary events in nature. They furnish a very innocent amusement, by indulging the natural taste of mankind for the marvellous, and I will readily allow that they sometimes throw light on the laws of nature, in her ordinary course of proceeding. I only mean to censure this extravagant attachment to prodigies, when it makes us neglect enquiries of more general utility to mankind. It deserves likewise to be remarked, that all great lovers of the marvellous are remarkable credulous, and have such heated imaginations, that every prodigy swells under their description.

There is another kind of facts, which, published to the world by themselves, answer no end but to fill up a volume; I mean, facts which are universally known, and of which everyone, who has [124] the use of his eyes, may collect as great a number as he pleases. Medicine is oppressed with cases of this kind; single cases that have no useful point of view, that neither tend to distinguish the disease more exactly from others which it resembles, nor to illustrate its remote and proximate causes, nor to establish its prognostic symptoms, nor to point out any better method of treating it than the common one, nor to ascertain the effects of any remedy.

The present fashionable taste for natural history, regards it more as an object of curiosity, than as the basis of a sound philosophy, subservient to medicine, agriculture, and the other useful arts. Every natural production is not only tediously described, but painted with the utmost elegance. We now have a superb folio which contains nothing but the natural history of a frog, in which that animal is beautifully painted in a great variety of attitudes. [125] We have folios executed in the same manner, which contain the natural history of the Danube, and of every thing which is found upon its banks. In this unmeaning taste of accumulating natural history, it is evident that books may be multiplied beyond number, without bringing any accession of useful knowledge; that the expence of procuring them must exceed what any private fortune can afford; and that they can only be admitted into public libraries, where they be exhibited like any other shewy piece of furniture.

XII. The advancement of the sciences has been much retarded, in consequence of the following causes.

1. Inattention to their ultimate end.

One of the chief causes that has obstructed the advancement of the sciences, has been an inattention to the principal end which should be kept in view in their cultivation: the end I mean is public [126] utility, or what contributes to the convenience and happiness of life. Instead of attending to this, most men have no other object in the pursuit of knowledge than to gratify a transient curiosity, or to give a variety to their amusements, or to serve the purpose of vanity and ostentation, or to gain a subsistence in the profession they live by. Perhaps there never was a science that has suffered so much as medicine, by the neglect of its ultimate end and purpose, which, as I before observed, was to preserve health, to prolong life, and to cure diseases. It has, indeed, made the slowest progress of any of the useful and practical arts; not surely from any deficiency of genius in physicians, but rather from exhuberance, or misapplication of genius; nor yet from want of erudition, for no profession can boast of more men eminent for every branch of useful and polite literature, than physic. They have not only cultivated, with greatest success, every science intimately connected with their own profession, such as anatomy, [127] botany, chemistry, and the various branches of natural history, but have often distinguished themselves as poets, mathematicians, and philosophers. Yet how few physicians can we name, who either by their genius or industry, have advanced the practical part of their own profession; how many, on the contrary, could we name, who have corrupted it, by the sportings of their own imaginations, dignified with the name of philosophy; and even checked the slow improvement, which time must naturally bring to every art founded on observation and experience. The reasons why medicine has made such slow progress, in comparison of the other practical arts, may be partly referred to the difficulty and intricacy of the art itself, and partly to some peculiar disadvantages which the profession lies under, which I shall afterwards endeavour to explain.

2. There is a certain metaphysical subtlety, which is not only useless in our [128] enquiries into nature, but does real mischief, by giving ingenuity and industry a wrong direction. This involved all science, for many ages, in darkness and endless controversies. It was carried to the greatest length by the schoolmen, many of whom having great acuteness, abundance of leisure, from their retired monastic life, little acquaintance with the best authors, and still less with the works of nature, spun out of a small quantity of matter, those cobwebs of learning, admirable indeed for the fineness of the thread, but of no substance or utility. As their writings consisted of nominal subtleties, and a play of words; as they occasioned perpetual wranglings, and led to no useful consequences, the wiser part of mankind became quite weary of them, and now the old school-philosophy has fallen into universal contempt. This philosophy corrupted no science more than medicine. From the days of Galen, till towards the end of the last century, all the institutions of physic were not [129] only filled with the chimerical doctrine of elements and temperaments, but such questions as these, whether the procuring of health be the design or end of medi-

cine; whether disease is a quality or relation; and innumerable trifles of a like kind. They are generally disputes about words; and whenever the terms are defined, the controversy is at an end. It is a melancholy thing, to reflect on the industry, erudition, and genius, so copiously displayed in the writings of the old physicians, and so superior to what is generally met with, among those of the present age: at the same time, to find them wasted in such disputes as disgrace the human understanding, and was employed in corrupting and embarrassing an art, that requires to bring it to perfection, rather attentive and sagacious observation, and a clear and solid judgment, than great metaphysical acuteness.

A useless subtlety may be displayed in two ways, either in the prosecution of [130] enquiries of no importance, but difficult investigation, or by treating important subjects in a way that leads only to fruitless speculation and controversy. We have examples of the first in the old school-logic, and in most metaphysical disquisitions, ancient or modern. I acknowledge the usefulness of such disquisitions, considered as an exercise for young minds. They may sharpen the invention, strengthen and improve the reasoning faculty, and communicate a power of fixt attention and nice discrimination; but when long dwelt upon, they withdraw the attention from the study of nature and the practical arts, and beget a habit of wrangling upon every subject, extremely disagreeable in conversation; because it tends to confound than to convince, and seems a contention rather for victory than truth. The habitual practice of balancing things, with a minute exactness and finical precision, is unfavourable to the enlarged views of genius, the advancement of the sciences, and the [131] successful management of business in private life. These require only an attention to probabilities, to leading principles, and the great outlines of objects, a quick discernment where the greatest probability of success lies, and habits of acting, in consequence of this, with facility and vigour.

Important subjects of enquiry are treated in a manner that leads only to fruitless speculation and controversy, when we waste our labour in a minute discussion of supposed necessary preliminaries, and points essentially connected with them, though, in reality, they have no connection at all, or a very remote one. It is the same useless labour, when we plunge at once too deeply into our subject, and attempt the investigation of causes, either beyond our reach, or such as, if known, could lead us to no useful consequences. Thus philosophers, before Sir Isaac Newton's time, were often attempting to explain the cause of gra-[132]vity. That great man contented himself with investigating the laws according to which it acts, and only proposes a suspicion of its cause in the modest form of a query. The laws according to which gravity, magnetism, and electricity act, are a proper subject of enquiry; because they are within our reach, and because the knowledge of them leads to the most useful consequences: But their causes will probably ever escape our deepest researches, nor, perhaps, could the discovery be of the least utility. The law of motion between the soul and body is one of the most important enquiries in medicine; but the enquiry into the nature of this union, is equally obscure and unnecessary.

3. There is another species of useless subtlety, which consists in an extremely scrupulous exactness, in regard to arrangement and method. These should, without doubt, be particularly attended to in treating of any subject, but are pe-[133]culiarly necessary in all the departments of natural history. The proper distribution of plants, and other productions, into their several orders, genera, and species, is a great assistance to the memory, and leads to important general observations, in relation to their virtues. But a compleat classification is a matter of the greatest difficulty, and can never be attained, without an exact knowledge of all the particulars proposed to be classed. It may be attempted upon different principles, as is the case with the various systems of botany, and though one of them may, upon the whole, be more perfect than the rest, yet each of them may have its own peculiar advantages. In the same manner, diseases may be classed according to their symptoms, their remote or proximate causes; and in various other ways, all of them very imperfect, but each having its own advantages and disadvantages. Neither is it possible for human ingenuity to remove this imperfection, till such time as [134] the knowledge of particular diseases, and the science of medicine, is rendered perfect. It is evident, therefore, that this subject of arrangement presents an ample field for disputes, where much ingenuity may be displayed; though, in fact, it is only in a specious kind of trifling. In this manner the attention is diverted from the study of diseases, and the most successful methods of treating them, to a fruitless speculation about the order in which they should be treated. I only call them fruitless, so far as they waste too much of that time and attention which might be more usefully employed. If we carry our studies in natural history no farther than to a just arrangement, what we have learned is of no more consequence, than the knowledge of a Greek grammar, and of all the words in a Greek dictionary, would be to one who was never to look into a Greek writer. I speak of natural history with real regret, as I see its principal purpose too much neglected. I see [135] it studied rather as a matter of curiosity, or as

furnishing subjects of ingenious speculation, than as subservient to real utility. It is of little importance to settle the genus and species of plants, in comparison of ascertaining their uses; yet the one subject has been attended to very closely, the other has been worse than neglected; it has been corrupted by a multitude of false facts, especially in what relates to medicine. Much pains have been taken to place those worms that infest the human body in their proper ranks, and to examine their anatomical structure with the greatest accuracy; but little proportionable care has been taken to fix the certain symptoms of their existence in the body, the effects they produce there, and the most effectual method of destroying them. I cannot, however, omit, on this occasion, doing justice to the merits of that great man, Dr. Linnæus, who has displayed so original a genius, in reducing all the subjects of natural history into so per-[136]fect and beautiful a system. Dr. Linnæus did not stop here; he has shewn the most enlarged spirit of observation, in applying natural history to the useful purposes of life, particularly agriculture and medicine.

4. The advancement of the sciences has been much retarded by a weak credulity of those who have cultivated them. This credulity discovers itself, in regard to particular facts, in a fond belief in the powers of certain delusive arts, in a bigotted attachment to some great names in the learned word, or in a superstitious veneration for antiquity.

(a) An easiness of belief, in regard to facts, by admitting them upon weak authority, has corrupted every branch of natural knowledge, but none of them so much as medicine. Facts depending upon the animal œconomy, must be difficult to ascertain; because they are subjected to all varieties, which they [137] are exposed to from a thousand nameless causes. A heated imagination, therefore, may easily fancify or magnify them, and fraud may easily forge or counterfeit them, when, at the same time, it is very difficult to detect the error. Hence our accounts of the effects of remedies still remain full of uncertainties and falsehoods; while many other branches of natural history, particularly chemistry, have of late been exceedingly well cleared of false facts. Medicine suffers much more from this cause than from fanciful theories. The weakness of a theory is easily detected. The clear understanding of one sensible man is sufficient of this. But there is frequently required the united labours of many to make a separation between facts that are fully and candidly represented, and such as are false or exaggerated; nor can it be done till such time as an opportunity offers of repeating the observation or experiment, perhaps at the risque of a patient's life. I do not mean to insinuate here, that no facts [138] should be admitted into natural history, or medicine, but such as are thoroughly established. I mean only to shew the impropriety of jumbling uncertain reports and undoubted truths, without making a proper distinction between them. Whatever is asserted to be a fact, let it appear ever so extraordinary, and though the authority for it be but slender, yet deserves to be recorded, until an opportunity offers of ascertaining its truth; as nothing shews more ignorance of nature, or more contemptible self-sufficiency, than the rejection of facts, merely because we cannot account for them.

(b) A fond belief in the powers of certain delusive arts, particularly astrology, natural magic, and alchemy, has greatly retarded the progress of knowledge, by engrossing the attention of many of the finest geniuses which the world has ever produced, and by introducing, into medicine especially, a multitude of false [139] facts, founded on the grossest superstition and delusion. These arts, which promised to be of infinite use in life, laid such fast hold on the imagination, that no power of reason was able to free men from their enchantment. At the same time, they have accidentally given rise to some important discoveries, and would furnish some excellent materials for the history of the human imagination.

(c) A bigotted attachment to certain great names in the learned world, has done remarkable mischief to science. The history of philosophy exhibits to us, from time to time, some man of distinguished ingenuity who has erected a system. This system has been universally adopted for a few years. Learned men have commented upon it; some have explained it with the most unsupportable diffuseness; others have abridged it. In the mean time, none of those authors rose higher than their source; few of them so high. In the [140] succession of a few years another original genius has started up, exposed the weakness of his predecessor's system, and established another in its stead. This, after having the like honours paid to it by commentators, has sunk, in its turn, into contempt and oblivion. This has been the fate of medicine, from the days of Hippocrates down to the present time, when there appears to be a general disposition arising to throw off the shackles of authority, to appeal to nature in matters of fact, and to assert the right of private judgment in matters of opinion and reasoning. I do not mean to insinuate the possibility of every individual's thinking for himself in these matters of science. Nature never intended the bulk of mankind either to think for themselves. I only mean to regret, that men, blessed with superior talents, should crouch to an authority they ought to have controuled, and should tamely yield their assent to doctrines, which a little exercise [141] of their own judgments would have shewn to be ridiculous.

(d) Another obstacle to the improvement of science, similar to the former, has been a blind and superstitious veneration for antiquity. It is inconceivable to those who are only acquainted with the present state of the learned world, and with the free spirit of enquiry in matters of science that now prevails, to what an absurd height this attachment to antiquity was formerly carried; how much it has cramped the efforts of genius, and retarded the progress of knowledge. Upon the revival of learning, all the ingenious men in Europe were employed in recovering, translating, and commenting on the remains of antiquity, which had escaped the ravages of time and barbarism, and lain for many centuries buried in the cells of monks. The world is infinitely obliged to the labours of these restorers of learning, which [142] quickly dispelled the darkness and ignorance, which for twelve centuries had overspread all Europe. The immediate effects produced by the recovery of the ancient writers, shewed very clearly in what their principal excellency consisted. All the fine arts, painting, sculpture, architecture, rose very speedily to an amazing degree of perfection. Purity of language, and an elegant simplicity of composition, especially in poetry and history, were particularly studied; but natural history and natural philosophy remained miserably neglected. The reason was plainly this; in all works of taste and imagination, in poetry, in eloquence, in simplicity, correctness, and elegance of composition, the ancients possessed an excellence hitherto unrivalled. In abstract mathematics, likewise, they will ever remain as standards of that clearness and precision, which should be the peculiar characteristics of mathematical reasoning. But in natural history, and in natural philosophy, they were not equally [143] successful. This was owing partly to their not having bestowed sufficient attention on those subjects, and partly to those sciences depending for their advancement, not so much on the genius of one man, as on the accumulated labours of many. Thus a Homer, an Appelles, a Praxiteles, or a Demosthenes, may have carried poetry, painting, sculpture, or eloquence, as high, or higher, than any who have succeeded them; because when these men died, their arts, in a great measure, died with them. But in natural history and natural philosophy, the case is widely different; because every man, who applies to any branch of these studies, may avail himself of all the labours and improvements of his predecessors. As the sciences, then, at the revival of learning, were in a low state; and as little light was thrown on them by the writings of the ancients, they continued to lie in a great measure neglected, till towards the middle of the last century; most men of learning and ingenuity, before that time, devoting all [144] their attention to theological studies, the fine arts, and the different branches of polite literature.

The same warm admiration of antiquity which prevailed in other subjects at the restoration of learning, attached physicians to the ancient writers in their own profession. It had been happy for mankind if, instead of a blind and stupid admiration of Hippocrates, justly stiled the father and founder of medicine, they had imbibed some share of his enlarged spirit for observation. Hippocrates will always be held in the highest esteem, for his accurate and faithful description of diseases, for his candour, his good sense, and the simple elegance of his stile. But, instead of prosecuting his plan, and building on the foundation he had laid, his successors employed their time in commenting on his works. Galen began with writing the most voluminous commentaries on what he reckoned the genuine productions of Hippocrates, in which he endeavours to reconcile all [145] his real and seeming contradictions, and to prove the truth of his observations by a variety of arguments, not drawn from on his own extensive experience, but from the Aristotelian philosophy; some of them, indeed, subtle and ingenious, but for the most part trifling and sophistical. This manner of writing commentaries on a book consisting of observations, is extremely absurd. The first and capital enquiry here ought to be into the truth of the facts. Till these are established by corresponding observations, it is a waste of time and labour, to attempt an explanation of the causes of them. Hippocrates has left us a number of excellent observations; some that are found to be true only in certain cases, and under certain limitations; some that are false and ridiculous; and a great number that seem curious and important, which not one of his numerous commentators has taken the trouble to enquire, whether they were true or false. Every one of them has, after the example of Galen, attempted to prove the truth of his ob-[146]servations, not by similar observations of their own, but by hypothetical reasoning, drawn from the prevailing philosophy of the particular times they lived in. Thus the noble foundation of observations begun by Hippocrates, and the example he has set of faithful and accurate description, have, in a great measure, been neglected, while physicians, in all ages, have been solicited to shelter all their theories of the most opposite kinds, all their sense and nonsense, under his authority, which the brevity and obscurity of many passages of his writings rendered an easy matter. Not only his observations, but his opinions, (of which indeed he was very sparing,) were adduced till very lately, to quash the authority of facts, which appealed for their truth to the experience of every man of candour and common sense; so that a physician, in writing his own observations, found himself under a sort of necessity to shew that they agreed with those of Hippocrates, at least that they [147] did not contradict them. The effect of this was, that the truth of Nature was often warped and perverted; and in order to make it appear correspondent to the sentiments of Hippocrates, or even with the more worthless authority of Galen. This introduced a corruption into the very source and spring of all solid knowledge in medicine; and, at the same time, encouraged a pompous and useless display of learning in writing on medical subjects, that wasted the time and tired the reader, who wanted to know what nature said, not what Hippocrates and Galen thought, in medicine. Neither is this pedantry yet extinct in Europe; there being few medical books wrote in some parts of it, but what are not stuffed with numerous quotations from the ancients, containing some very trite observations, that answer no other purpose, but to make a parade of the author's erudition, and to swell the volume. [148]

5. Another obstruction to the progress of science, the very reverse of the former, has been a fond attachment to novelty. This proceeds, partly from a natural principle in the human mind, which is gratified, in a certain degree, with whatever is new, independent of other consideration, partly from an anxiety to discover truth upon an interesting subject, which makes us often grasp a shadow for the substance, and partly from a disposition to believe whatever we earnestly wish to be true. The uncertainty of the methods of cure, in many diseases, makes patients, and sometimes physicians, very eagerly adopt any new method, that promises a more effectual and speedy cure. This is the cause of that universal propensity to give credit to the extravagant and exaggerated accounts of the effects of nostrums and quack medicines. These are recommended to a patient, with an assurance of their infallibility, which no physician, who has regard to either honour or prudence, can in any case venture to [149] give. From the same cause we have seen, in our own times, many remedies loaded with encomiums, to which they had no title, becoming fashionable, for some years, in almost every disorder, and then sinking into neglect, such as, cold water, crude mercury, soap, tar-water, lime-water, sea-water, Dr. Ward's[4] medicines, and now many of the class of poisons. During the reign of these medicines, the public was amused with a belief that they were all infallible, in the cure of almost all chronic disorders;

and when time discovered the folly of this expectation, they were, with equal folly, laid, in a great measure, aside; as if a medicine could not be useful in the cure of some disease, because it was not infallible in the cure of all. This attachment, however, to novelty, is not such a bar to improvement, as a superstitious veneration for antiquity. The former, from time to time, is bringing new accessions to knowledge; the latter keeps the active powers of the mind suspended [150] in a stupid admiration of what, perhaps, was of some value in the infancy of science, but what is now universally known to every smatterer in learning. A physician of sagacity and coolness may derive great advantages to his art, by these temporary intoxications of the public, in regard to such remedies, which he sees, but cannot prevent; as they give him an opportunity of ascertaining the effects of some medicines, by allowing him to exhibit them in larger doses, and during a greater length of time, than patients would otherwise be persuaded to make a trial of. The passion for novelty is particularly excusable in medicine; because it is natural for us to be pleased with what seems, not only to bring an accession to our stock of knowledge, but to communicate some useful discovery. But the discovery of our mistake is not an equally natural force of pleasure; because this makes no positive addition to our knowledge, but on the [151] contrary, deprives us of what we thought we have once gained.

6. The hasty reduction of any science into a system, apparently full and perfect in all its parts, while, in reality, these parts are ill filled up and erroneous, is an effectual bar to its farther improvement. The intention of these systems is to place a science in such a light as may produce it most credit. It is, therefore, delivered in a pompous and magisterial manner, so as to gain belief without examination; and hence a science descends in the persons of master and scholar, not of inventor and improver. Men are generally attached to systems, because they free them from the impatience of doubting, and promise them fixed and certain principles, on which their minds may securely rest; and teachers find it contributes both to their interest and reputation, to reduce the sciences into systems, as seemingly compleat as possible. A man who ap-[152]pears perfectly well acquainted with the principles of a science, and who seems to entertain no doubt of their soundness, makes a more shewy appearance, than one who doubts, and fairly owns that he does so. The bulk of mankind are not judges of the merit of men of science, and are ready enough to allow them that the consequence, which they take to themselves, if vanity does not make them greatly overshoot their mark. I have before endeavoured to

shew the propriety of prosecuting enquiries into nature upon a regular and methodical plan; in teaching a science, it is equally necessary to proceed upon a particular plan of arrangement. But, till such time as all the facts, and all the principles, included in a science are perfectly established, it is impossible to reduce it into the form of a compleat system; and there are many circumstances relating to arrangement, which, in the mean time, must remain undetermined. It is, therefore, sometimes better to use the loose [153] aphoristical manner, than to attempt an order, where there are no certain principles to lead to it. It has been the fate of medicine to suffer, in a very particular manner, from this unfortunate disease of system-making. It has fallen, at different times, into the hands of Galenists, Chemists, Cartesians, Mathematicians, Stahlians, and some other sects compounded of these; each of whom have moulded the whole sciences of medicine into a system, seemingly compleat in all its parts. It has been tinctured with mystical divinity, astrology, and all the subtleties of school philosophy, according to the different attachments of physicians to those studies. But, notwithstanding the load of learned rubbish with which it has been encumbered by systems, a physician of genius, and solid judgment, will be able to draw from them some useful information, although in view of compleat systems, he heartily despises them all. [154]

7. The last impediment I shall mention to the progress of science in general, has been too great attention to purity and elegance of language, on the one hand; and, on the other, an affected obscurity and intricacy of style. In works of taste, and addresses to the passions, a language highly ornamented may be very proper; elegance, sublimity, pathos, are there in their proper place. But the language in which science is to be communicated, should be simple, perspicuous, and divested of all flowery and artificial ornaments. Original writers, who have new ideas to communicate, are often obliged to use new words and phrases, in order to convey their meaning more distinctly and forcibly; which surely they, and they only, have a right to do, provided they clearly define them. An affected obscurity and intricacy of style is now, in a great measure, banished from the learned world. The use of technical terms, where others equally clear and expressive can be found, is regarded as [155] pedantry, or a cloak to conceal ignorance. This censure may sometimes be carried too far, but in general it is well founded. That learned jargon, which so long disgraced philosophy, was introduced for the illiberal principles of vanity, or for the still more unworthy purposes of shutting up the avenues of science, from all who did not live by it as a trade. But it evidently hinders the advance-

ment of science, when men attend more to words than things, whether it be in an affected and pedantic display of learning, or in a scrupulous regard to purity of diction, or elegance of composition.

XIII. Let me take this opportunity of recommending to your serious study the writings of Lord Bacon, who possessed, perhaps, a more enlarged and piercing genius, than any man who ever existed. He has explained the proper method of acquiring knowledge, and promoting science, with incomparable judgment and perspicuity. He has likewise left us [156] some beautiful specimens of true philosophical induction, particularly in his History of the Winds. This, and some other of his essays in natural history, are to be considered in no other light, than as specimens of his method of carrying on enquiries into nature. The facts they contain are not to be depended on: he was obliged to take such as were then generally believed, which, whether true or false, equally served the purpose of illustrating his method. He uses a language peculiar to himself: it is, beyond any other, the language fitted for science and philosophy, copious, clear, manly, and admirably expressive; but, unfortunately, incompatible with the affected delicacy of modern English, where he has become more feeble, in proportion as he has become more smooth and polished.

XIV. I have thus endeavoured to explain some of the principal causes that have obstructed the progress of science in general; and, where it was necessary, [157] have applied my observations particularly to physic. I thought it necessary to explain to you my general sentiments, in relation to the improvement of knowledge; because it gave me an opportunity of communicating my leading principles in the science of medicine. But, before I conclude the subject, a regard to truth and candour obliges me to take notice of some peculiar disadvantages under which medicine labours, and which in particular have retarded its progress. This I do not from any disposition to find fault, not from a desire to expose the weakness of a profession, the honour of which my inclination, and many ties, lead me to support; but merely with a view to put you on your guard against certain errors and inconveniencies, which you might otherwise be exposed to.

The peculiar disadvantages under which medicine has laboured, have arisen from the manner in which it has been usually taught, and from its having been [158] confined to a set of men who live by it as a profession.

1. In the first place, the general method of conducting education, in universities where medicine is taught, does not seem so well calculated to

advance science, as to diffuse it; not so well fitted to promote particular arts, as to communicate general principles. Those who teach the science, often lay various nets for the understandings of their students. Sometimes with the laudable view of engaging and fixing their attention; sometimes with a desire to stamp a dignity on their own characters, by pretensions to discoveries, by the triumph of confutation, the ostentation of learning, or the mask of obscurity. For the conveniency of teaching medicine, it has been usual, in most universities, to lay down general doctrines and principles, relating to entire classes of diseases and remedies, so far only as they serve to illustrate those prin-[159]ciples, or as they are clearly deducible from them. But the natural and genuine method of advancing a science is the reverse of the former, where we proceed from particular facts to establish general principles. Though, on a superficial view, it does not seem a matter of great consequence, in what particular way the knowledge of medicine is acquired; yet it will appear, on a nearer view, to have an important influence on a physician's future character and studies. Medicine, as usually taught in colleges, instead of being represented as an art, imperfect in its most material parts; instead of having its deficiencies pointed out, with a view to their being supplied, is digested into a regular and seemingly perfect system. In this view it is beheld by the young student, who embraces theories, with the same facility and unsuspecting confidence as he would do facts; he thinks he understands the causes of all diseases, and the manner of operation of remedies; his mind is at ease in having always sure [160] and fixt principles to rest on. In the mean time, the art has little chance to acquire improvement from him, as he scarcely supposes it stands in need of any. When a patient dies, he is quite satisfied every thing was done for him that art could do. It is difficult and painful for men to give up favourite opinions, the children of their youth; to sink from a state of security and confidence, into one of suspense and scepticism. Accordingly, few physicians change either the principles or the practice they first set out with. We have some remarkable examples of genius in physic, writing systems of practice, early in life, who have arrived at a very different old age, greatly admired for their capacity, and possessed of the most extensive practice; and though their systems had gone through many editions, yet there has been no material alteration of the last from the first: which affords a strong proof of the faithful attachment to their first ideas [161] and principles. Yet any person unacquainted with the history of physic, would naturally suppose that a physician, of accurate observation and extensive practice, should, in the course of a long life, have made such an addition to his stock of knowledge, as must necessarily have rendered his last performances of infinitely more value than his first; as must have confirmed him in some opinions, of which he was formerly doubtful; but discovered to him the folly or uncertainty of many more, whose truth, in his younger days, he had thought perfectly established. It we now enquire into the effects produced on the mind, by acquiring knowledge, in the slow method of induction, from observation and experiments, we shall find them very different. The mind here gains a habit of close attention to facts, having nothing else to trust to; slow in forming principles from these facts, and diffident of them when formed, instead of being assuming and dogmatical, becomes modest and sceptical. [162] A physician, whose knowledge has been formed in this manner, never loses a patient but he secretly laments his own ignorance of the proper means of having saved him, which he is always more ready to blame, than the incurableness of the disease itself. There are many diseases, which no physician has yet been able to cure; but it does not follow from this, that all these diseases are absolutely incurable. There are so very few diseases that can be pronounced, in their own nature, incurable, that I should wish you to annex no other idea to the phrase, incurable disease, but the idea of a disease which you do not know how to cure. How many patients have been dismissed from different hospitals, as incurables, who yet have recovered perfect health, sometimes by the efforts of unassisted nature, sometimes by very simple and safe remedies, and sometimes by the random and desparate prescriptions of ignorant quacks? To pronounce diseases incurable, is to establish indolence [163] and carelessness, as it were by law, and to screen ignorance from reproach. This diffidence of our own knowledge naturally stimulates us to improve it, not only from a love of the science, but from a principle of conscience and humanity. We own, that this philosophical spirit, if it is not united with great strength of mind, may be very detrimental to a physician, by making him timid and fluctuating in his practice: but though true philosophy leads to diffidence and caution, in forming principles, yet, when there is occasion to act, it shews how necessary it is to have a quickness in perceiving where the greatest probability of success lies, to be decisive in forming a resolution, and steady in putting it into execution. As every professor, of an enlarged mind must be sensible of the inconveniences that attend the usual method of teaching, he will guard against it by every method in his power, particularly by pointing out all the deficiencies in his system, and by promoting a spirit of free

[164] enquiry among his students, and an absolute contempt of the authority of all great names, in every thing but matters of fact. In these their authority must be submitted to, unless there be reason to doubt their integrity, or suspect their credulity. I throw out these observations with freedom from this place, where I am sure I cannot be misunderstood. In some universities in Europe, a little more caution might be expected; but I am well acquainted with the liberal spirit that breathes in this university, in every department of science, and in none more than in all the branches of medicine. But there are none of my obligations to it, which I remember with more gratitude, than the acquisition of that freedom of enquiry, which then distinguished it, and which so eminently distinguishes it at this time. Let me take this opportunity of doing justice to the merit of several gentlemen, who have, within these few years, done honour to this medical college, by their inaugural dissertations. [165] In these, several important investigations have been carried on, by a set of accurate and well-conducted experiments, under the direction of my learned and ingenious colleagues, particularly Dr. Cullen and Dr. Monro, which really tend to the advancement of science. This method of giving a specimen of a young physician's genius, is attended with so many advantages, so creditable to himself, and so useful to the public, that I should be extremely sorry to see it fall again into disuetude.

2. But to return to my subject. I would observe, in the second place, that the confinement of the practice of physic, entirely to a class of men, who have no other method of subsistence, is unfavourable to the progress of the art; because the spirit of improvement is often checked by, and is sometimes incompatible with, the necessary attention to private interest. Physicians are neither better nor worse than the rest of mankind. They are in-[166]fluenced by the same general motives of action. A physician, when he sets out in life, quickly perceives that the knowledge most necessary to procure him a subsistence, is not mere knowledge of his profession. What he finds more essential to that purpose, are the various arts of insinuation, and the arts of deceiving mankind into a high opinion of his undertaking, by an appearance of solemnity and importance in his deportment; views very different from those of genius and science. He can with difficulty find a patron to his real merit; because none are judges of it, but a few of his own profession, whose interest it is to have it concealed. If he attempts to shew them the weakness of the fashionable system, or to introduce any alteration in the practice, the whole faculty are alarmed; their vanity is piqued, in having opinions which they thought

perfectly established, brought into question, and exposed by a young man; and their interest is evidently concerned, [167] to crush him as soon as possible. In the mean time, the effect of every deviation, which he makes from the common practice, is anxiously watched, all his prescriptions must remain upon the apothecary's file, to rise in judgment against him; and upon any miscarriage, the outcry is raised and propagated with the utmost malignity. Now, as physic is yet but a conjectural art, no improvement can ever be attempted in it, without some risque of bad success; and the opportunities of misapprehension, not easily to be refuted, are so many, that few will venture the experiment. Sometimes, it happens, that a man of spirit rises above all these difficulties; but it happens, unfortunately, that this sort of boldness does not so often accompany literary merit, as diffidence and want of resolution. By what is said above, no reflection is intended to be thrown on mankind, as if they were natural enemies to superior merit. In the pursuits of interest, and of ambition, one man [168] does not hate another, because he is engaged in the same race; he desires him only to keep his proper distance behind, and he will wish him extremely well. Neither is it meant, while we describe the characteristical spirit of a class of men, to include all individuals of a profession. There is a conspicuous dignity, that generally accompanies genius, which renders those who possess it equally superior to suggestions of envy, and all the low arts of dissimulation.

But, not to insist further on arguments that shew, that no considerable improvement in the art of medicine can be expected from physicians, while they are on the present footing, I shall only observe as a fact, that it appears from the history of medicine, that the improvements in it were seldom owing to those physicians, who valued themselves upon being regular, systematic, rational practitioners; nay, what is more extraordinary, they have been opposed by them with [169] great keenness and acrimony, and seldom adopted till after a very long struggle. We could give instances of this, in many of the improvements of modern practice, particularly in the case of blisters, opiates, Peruvian bark, antimony, mercury, and all the powerful chemical remedies; the inventors or introducers of these, from the days of Paracelsus down to Dr. Ward, have been held by the faculty in contempt and detestation. The discoveries of those men who were not regular physicians, have not been always examined with that candid impartiality, which their importance and success required; yet from such men very useful discoveries may sometimes be expected. Quacks have advantages, in not being fettered like other physicians; as they seldom can suffer much, either in their interest or reputation, from the bad success of their experiments. But they have another great advantage above all regular physicians: they have much more extensive practice. Dr. Ward has prescribed for [170] more dropsies in a week, than any physician in Europe could do in a twelvemonth. I allow that the ignorance, carelessness, and wrongheadedness of most of that tribe makes them profit but little, in proportion to what might have been expected, from so extensive a practice; and I allow, that there is but little candour, or common honesty, in most of their accounts of cures. But it is a physician's business, to search for knowledge in his profession from all sources, however impure and contemptible; and he may avail himself of that enlarged experience, which an empiric cannot, nor will not, turn to account. It was from strolling chemists, and the lowest artificers, and not from the schools of philosophy, that Mr. Boyle drew that large and useful collection of facts, with which he has enriched philosophy. A strange fate seems, in many instances, to influence mankind. They are tenacious and jealous of their liberty and property: they still continue to clamour against [171] priestcraft, and the authority claimed by priests over their consciences, even now, when priests are living quietly and inoffensively, and not troubling themselves about any bodies consciences: yet they have trusted their health and lives into the hands of a class of men, without making any enquiry how they acquitted themselves of so important a charge. The science of physic has been sometimes advancing, sometimes declining; it has been subjected to the fate of different systems of philosophy that have prevailed, besides being sometimes disgraced by peculiar follies of its own; its only genuine source, observation and experiment, has been corrupted by fraud, credulity, and a heated imagination, while men of genius and learning, because they were not physicians, have kept at a distance, as if it had been a matter in which they were not interested. I cannot however, but observe, that the same manly and liberal spirit of enquiry, which has enlightened every other branch of natural knowledge, [172] begins to find its way into medicine; that the tyranny of authority and system declines apace; and that there is a fair prospect of the science being rebuilt on the more solid basis of nature, on facts, and an accurate induction from facts. It is said, by those who want to shew the propriety of confining the study of physic entirely to a class of men who live by it as a profession, that the science is so complicated, that it requires the whole of a person's time and attention to understand it. The little progress it has made, notwithstanding the labours of so many ingenious and learned men, entirely devoted to its cultivation, is adduced as a proof of its difficulty and intricacy. It is said, that if people were encouraged to study physic, who are not regularly bred to it, and who do not intend to practice it as a trade, that quacks would be multiplied, and that patients would lose that confidence and implicit faith in the physician, which is as convenient for their own sakes, as well as his. These [173] reasons have appeared so powerful to the medical faculty, that they have watched with the most jealous eye, over all intruders; and have usually treated them with great abuse and ridicule, even when it was apparent, that the intrusion was only owing to motives of humanity. It would not be candid to ascribe this to any sordid views: enlarged knowledge produces a liberal and unsuspicious spirit, and there is no profession can boast of more men of learning, ingenuity, and genteel education, than physic. But I must take the liberty to observe, that the difficulties which a gentleman, not regularly bred to the profession, is to meet with in acquiring some share of medical knowledge, are exaggerated greatly. Medicine, considered in one point of view, is a science of infinite extent. I know not any other that presents so ample a field for the exertion of genius and industry; the purposes of it are of the utmost consequence to mankind, and it will require the united labours of many ages, [174] however well they may be conducted, to bring it to such a state of perfection, as may even entitle it to the name of a science, according to the old school definition of a science, "cognitio certa & evidens." If, again, we consider medicine in view of a profession, by which a gentlemen is to live, it is still difficult and complicated. A physician of spirit, who would wish to appear with dignity in his profession, must be acquainted with various branches of knowledge, which are rather ornamental than essential to the main ends of his art; although he will be able to make the separation in his own mind, between the liberal accomplishments that distinguish the gentleman and scholar, and that knowledge which is indispensibly requisite to his practising with any degree of credit or success. A private gentleman, who has a literary turn, and chooses to study medicine as a curious and interesting branch of natural history, but who does not propose to practise it as a trade, may consider it in a view different from either of these. [175] Such a one has no intention to prosecute medicine with a design to improve the science; and requires none of those ornamental accomplishments, which give importance to a physician in the eyes of the world. He wishes only for such a degree of knowledge, as may enable him to understand such books of merit, as have been written on the subject of physic, and to judge of the comparative merit of those men to whom he is to commit the important charge of his own health, and the health of those whom he is obliged, by the ties of nature and humanity, to take care of. This is a matter of no such great difficulty as it is pretended to be. It requires, indeed, such a knowledge of anatomy as in necessary to understand the animal œconomy, both in its sound and morbid state. It requires a knowledge of the principles of chemistry, and of the operation of remedies: it requires an acquaintance with diseases, and the usual method of treating them. This may be taught by any good system of practice. These systems are all [176] compilations, that differ from one another, rather in neatness and elegance, than in any thing material. The latest system often has the advantage, in containing all the fashionable prescriptions; but none of the facts which they contain can be entirely depended on, unless ascertained by personal experience; from which source alone, from an attentive observation of the sick, and the effects of what is prescribed, all solid knowledge on this subject must be derived. Next to his own experience, he will learn most from the conversation of a sagacious and candid physician, who will direct his studies, and communicate the result of his own observations. If ingenious men would devote half the time to the study of nature, which they give to the study of opinions, true philosophy would make a very rapid progress. If a gentleman has a turn for observation, the natural history of his own species is a more interesting subject, and presents a more ample field for the exertion of genius, [177] that the natural history of spiders and cockle-shells. If such men were to claim their right of enquiry into a subject that so nearly concerns them, the good effects on medicine would soon appear. They would have no separate interest from that of the art. They would detect and expose assuming ignorance, under the mask of gravity and importance, and would be the judges and patrons of modest merit. Cases very often occur where an ingenious physician sees his patient hastening to certain death; he knows a remedy that affords a probable prospect of saving his life, but it is uncommon, not agreeable to established orthodox system, and dangerous in its operation. Here is a dreadful dilemma. If he gives the remedy, and the patient dies, he is utterly ruined. The dunces, who are the most numerous in every profession, are always at war with genius, and watch its miscarriages with the most anxious and malignant eye. But in such a case, the encouragement and assured protection of [178] knowing and disinterested judges would animate a physician to do his duty. Such men, not having their understandings perverted in their youth by theories, unawed by authority, and unbiassed by interest, would canvas with freedom the most universally received principles in medicine, and expose the uncertainty of many of those maxims which a physician dares not seem to doubt of. Lord Bacon had as enlarged views in medicine, of its deficiencies, and of the proper method of supplying them, as perhaps any physician that ever wrote. Dr. Hales has been one of its greatest benefactors. Cornaro, a Venetian nobleman, when some years turned of fourscore, composed a little treatise on regimen, written with more candour, simplicity, and precision, than any thing I have seen on the subject. With more pleasure could I name Mr. Boyle on this occasion, had not his credulity lessened that esteem, which his diligence, genius, and many virtues, so well merited. It is said, that if the my-[179]steries of the art were thus to be laid open to those who were not regularly initiated, it would destroy the physician's authority, and that implicit faith which the sick ought to have in him, for their own sakes. But, in fact, his authority is controuled by all, except those who alone should have any title to controul it. All the midwives, nurses, and old women are physicians; and the dignity of the most stately of our faculty is often obliged to stoop to the follies and caprices of such people, who are sometimes of more consequence in making a physician's fortune, than all the merit he can possess. The only tame and believing patients are men of sense, who generally submit to their physician, whoever he is, with wonderful faith and patience. But if these men were to devote part of their time and studies to medicine, they could more effectually support the physician's authority against the encroachments of ignorant pretenders; and, at the same time that they were [180] conscious of the superiority of a physician of extensive learning and practice, yet they might occasionally suggest hints of great consequence to the ablest physician. We may here observe, that the same objections made against any person's pretending to judge of medical subjects, who has not been regularly bred to the profession, were formerly made against the reformers from Popery. Besides the Divine authority claimed by the church, it was said, that a set of men, who devoted their whole time and studies to so deep and complicated a subject as theology, were the only proper judges of whatever belonged to it; that calling their authority in question, was hurting the cause of religion, and bringing the sacerdotal character into contempt. Yet experience has shewn, that since the laity have asserted their right of enquiry into these subjects, theology, considered as a science, has been improved, the interests of real religion have been promoted, and the clergy have become a [181] more learned, more useful, and a more respectable body of men, than they ever were in the days of their greatest power and splendor.

By what I have said, I hope it will evidently appear, that I have no intention to lessen the dignity of a profession, which has always been considered as most honourable and important. But, I apprehend, this dignity is not to be supported by a narrow, selfish, corporation spirit, by a peculiar formality in dress and manners, or by affected airs of mystery and self-importance. The true dignity of physic is to be maintained by the superior learning and abilities of those who profess it, by the liberal manners of gentlemen, and by that openness and candour, which disdain all artifice, which invite a free inquiry, and which, by this means, boldly bid defiance to all that illiberal ridicule and abuse which medicine has been so much exposed to.

FINIS. [182]

JOHN GREGORY

LECTURES ON THE DUTIES AND QUALIFICATIONS OF A PHYSICIAN

Citation:
Gregory, John
Lectures on the Duties and Qualifications of a Physician
London: W. Strahan and T. Cadell, 1772

Title page carries this information on Gregory:
"Physician to His Majesty, and Professor of Medicine in the University of Edinburgh"

Comment: No changes have been made to the original text. In particular, original spellings and punctuation have been retained. Spellings of particular words may vary in the text; these variations have been retained. The goal has been to replicate the original text in its entirety. Page numbers appear in brackets at the end of each original page. '[n.p.]' indicates an unnumbered page break in the original.

Acknowledgment: This text was prepared from a copy of *Lectures* in the Blocker History of Medicine Collections in the Moody Medical Library of the University of Texas Medical Branch at Galveston. I am grateful to Inci Bowman, Ph.D., Curator of the Blocker Collections for her assistance.

Text:

LECTURES
ON THE
DUTIES AND QUALIFICATIONS
OF A
PHYSICIAN [N.P.]
LECTURES
ON THE
DUTIES AND QUALIFICATIONS
OF A
PHYSICIAN
By JOHN GREGORY, M.D.F.R.S.
Physician to His MAJESTY and Professor of Medicine
in the University of Edinburgh
A NEW EDITION, corrected and enlarged
LONDON: Printed for W. Strahan; and T. Cadell, in the Strand
MDCCLXXII [n.p.]

DEDICATION

TO
SIR JOHN PRINGLE, BART
PHYSICIAN TO HER MAJESTY

Sir,

WITH great pleasure I embrace this opportunity of giving you a public testimony of my sincere attachment. There is, besides, a peculiar propriety in addressing to you the following Lectures, intended for the use of the young students in physic, as it affords me a very proper occasion of pointing out to their imitation, a gentleman, whose honour and pro-[n.p.] bity, whose genius and learning, have done so much credit to the profession, and whose ardent zeal and unwearied labours have so much contributed to its advancement.

I am, with the sincerest respect and esteem,
Sir,
Your obliged and
Faithful Servant,

JOHN GREGORY
College of Edinburgh
May 19, 1772 [n.p.]

ADVERTISEMENT

The following Lectures have been read in the University of Edinburgh for several years past, and, as many transcripts of them were, from time to time, taken by my pupils, one of them found its way to the press in the negligent dress in which they were first exhibited. The Public, however, having been pleased to afford them a favourable reception even in that form, I thought it a piece of justice I owed to its candour, to give them a thorough revisal, and to make them, as far as I was able, more worthy of their acceptance. This I have now done. I hope they will be found of some use not only to students, but to the younger [n.p.] part of the Faculty; and that my sincere endeavours to promote the true interests of Physic, however ineffectual, will induce my Brethren to overlook any defects that, after all my care, may still be found in them. [n.p.]

errata [on an interleaved, partial page, between end of "Advertisement" and p. 1]

Page 7. _ line 16. *for* reflects, *read* reflect.
8. _ 13. *read* which is not to be purchased.
16. _ 24. *read* and resolution to act.
21. _ 21. *read* in order to conceal
38. _ 23. *read* but the following cause.
41. _ 21. *for* patience, *read* abilities.
46. _ 15. *read* or whether we should wait.
48. _ 8. *read* or charge it at the prime cost.
116. _ 2. *for* testimony, *read* evidence.
128. _ 8. *for* consider, *read* think.
148. _ 14. *read* of such.
149. _ 8. *dele* were.
155. _ 8. *for* two, *read* too.
158. _ 12. *read* described with a tiresome minuteness.
164. _ 23. *read* points which we fancy to be essentially connected with them.
184. _ 4. *read* together with some that are found.
207. _ 12. *read* suppose of a young man.
218. _ 9. *read* depreciate their conduct.
222. _ 13. *dele* the merits of the.
235. _ 8. *for* himself, *read* they.

[pagination numbering begins with this page]

LECTURE I

Utility and dignity of the medical art. – Reasons why physicians have been sometimes exposed to ridicule. – Requisites to form the character of a physician. – Opportunities which the profession of medicine gives for the exertion of genius, and for the exercise of humanity. – Enquiry into the duties and offices of a physician. – Division of the subject. – The genius, understanding, and temper required in a physician. – Difficulties attending the profession. – Command of temper, presence of mind, and resolution necessary. – Moral qualities. – Humanity. – Gentleness of manners. – Flexibility. – Particular tenderness due to nervous patients. – Frequent contrast between the manners of a physician when first setting out, and when established in [1] practice. – Obligations to discretion, secrecy and honour. – Temperance, sobriety. – Candour. – Openness to conviction.

The design of the professorship which I have the honour to hold in this university, is to explain the *practice of medicine*, by which I understand, the art of preserving health, of prolonging life, and of curing diseases. This is an art of great extent and importance; and for this all your former medical studies were intended to qualify you.

But, before I enter upon the particular business of this course, I shall, agreeable to custom, give some preliminary lectures, in which I shall lay before you some considerations, which though not strictly belonging to my subject, yet deserve the attention of all those who would practise medicine. – On this occasion I think it needless to dwell on the utility and dignity of the medical art. Its utility was never seriously called in ques-[2]tion; every man who suffers pain or sickness will very gratefully acknowledge the usefulness of an art which gives him relief. People may dispute, whether physic, on the whole, does more good or harm to mankind; just as they may dispute, whether the faculty of reason, considering how it is often perverted, really contributes to make human life more or less happy; whether a vigorous constitution and an independent fortune are blessings or curses to those who possess them; whether the arts and sciences in general have proved beneficial or detrimental to mankind. – Such questions afford opportunities for the display of eloquence, and for saying plausible and ingenious things; but still nobody doubts of the real and substantial advantages attending those acquisitions, if applied to their natural and proper uses. Much wit has, indeed, in all ages, been exerted upon our profession; but after all, we shall find that this ridicule has rather been employed against physicians than phy-[3]sick. There are some reasons for this sufficiently obvious. Physicians, considered as a body of men, who live by medicine as a profession, have an interest separate and distinct from the honour of the science. In pursuit of this interest, some have acted with candour, with honour, with the ingenuous and liberal manners of gentlemen. Conscious of their own worth, they disdained every artifice, and depended for success on their real merit. But such men are not the most numerous in any profession. Some impelled by necessity, some stimulated by vanity, and others anxious to conceal ignorance, have had recourse to various mean and unworthy arts, to raise their importance among the ignorant, who are always the most numerous part of mankind. Some of these arts have been an affectation of mystery in all their writings and conversations relating to their profession; an affectation of knowledge, inscrutable to all, except the adepts in the science; an air of perfect confidence in [4] their own skill and abilities; and a demeanour solemn, contemptuous, and highly expressive of self-sufficiency. These arts, however well they might succeed with the rest of mankind, could not escape the censure of the more judicious, nor elude the ridicule of men of wit and humour. The stage, in particular, has used freedom with the professors of the salutary art; but it is evident, that most of the satire is levelled against the particular notions, or manners of individuals, and not against the science itself.

Of the dignity of the profession I need say little. I suppose you are well satisfied that you have chosen a reputable one. Whatever may have been the pride or caprices of a few countries, it has generally been looked upon, and with good reason, as one of the most liberal. To excel in it requires a greater compass of learning than is necessary in any other. A knowledge of mathematicks, at least of the elementary parts of them, of na-[5]tural history, and natural philosophy, are essentially connected with it; as well as the sciences of anatomy, botany, and chemistry, which are indeed its very foundations. There are likewise some parts of knowledge, which, though not absolutely necessary to the successful practice of medicine, are yet so useful, that no physician, who has had a regular education, is found without them; such are, an acquaintance with the Latin, Greek, and French languages. If you add to this, that knowledge of men, and of manners, which a physician naturally and insensibly acquires by an extensive intercourse with mankind, I think it will evidently appear, that no profession requires a greater variety of liberal accomplishments than that of physick. This sufficiently establishes its dignity: I say, its dignity, if that is to be estimated by its real usefulness to mankind, and by the variety of talents necessary to practise it with success and reputation. [6]

We have indeed much reason to be pleased with the honourable point of view in which our profession is regarded in every part of the British dominions. They who have seen in how contemptible a light some of its branches are considered in other countries of Europe, will feel more sensibly the just regard paid to them here. One happy consequence, among many others, which results from this, is, that gentlemen of the best families, distinguished for their spirit and their genius, often apply to the study of medicine; and the liberal and ingenuous manners, generally found in men well born and genteelly educated, reflects [corrected to 'reflect' in errata] an additional dignity on the profession.

Besides the general consideration of the utility and dignity of the science of medicine, it may be considered in two different views. [7]

In the first place, as presenting a very ample field for the exertion of genius. – The great extent of the subject, and a variety of causes, which I shall afterwards endeavour to explain, have left it imperfect in many of its parts; and indeed their are some in it hitherto unexplored.

In the second place, medicine presents a no less extensive field for the exercise of humanity. A physician has numberless opportunities of giving that relief to distress, ['which is' added in errata] not to be purchased by the wealth of India. This, to a benevolent mind, must be one of the greatest pleasures. But, besides the good which a physician has it often in his power to do, in consequence of skill in his profession, there are many occasions that call for his assistance as a man, as a man who feels for the misfortunes of his fellow-creatures. In this respect he has many opportunities of displaying patience, good-nature, generosity, compas-[8]sion, and all the gentler virtues that do honour to human nature. The faculty has often been reproached with hardness of heart, occasioned, as is supposed, but their being so much conversant with human misery. I hope and believe the charge is unjust; for habit may beget a command of temper, and a seeming composure, which is often mistaken for absolute insensibility. But, by the way, I must observe, that, when this insensibility is real, it is an misfortune to a physician, as it deprives him of one of the most natural and powerful incitements to exert himself for the relief of his patient. On the other hand, a physician of too much sensibility may be rendered incapable of doing his duty from anxiety, and excess of sympathy, which cloud his understanding, depress his spirit, and prevent him from acting with that steadiness and vigour, upon which perhaps the life of his patient in a great measure depends. [9]

This naturally leads me to make some observations on the duties and office of a physician; a subject of great importance, but perhaps of so delicate a nature as makes it difficult for one of the profession to treat it with proper freedom. I shall, however, attempt to do it, without any reserve. The difficulty of treating this subject in such a manner as to give no offence, arises from hence, that medicine may be considered either as an art the most beneficial and important to mankind, or as a trade by which a considerable body of men gain their subsistence. These two views, though distinct, are far from being incompatible, though in fact they are too often made so. I shall endeavour to set this matter in such a light as may shew that the system of conduct in a physician, which tends most to the advancement of his art, is such as will most effectually maintain the true dignity and honour of the profession, and even promote the private interest of such of its members as are men of real ca-[10]pacity and merit. I am under less apprehension of discussing this subject before gentlemen at your time of life, than if you were further advanced in years. Youth indeed is the season when every sentiment of liberty, of generosity, and of candour, most easily find their way to the heart. If they do not reach it then, they never will afterwards. Age may improve the understanding by accessions of knowledge and experience; whilst at the same time that warmth of temper and imagination, which so often mislead the judgment, gradually abate. But it unfortunately happens that this very circumstance attending the decline of life, which in some respects improves the understanding, in others throws a damp upon genius, checks the ardent pursuit of science and truth, and shuts the heart against every manly, enlarged, and generous sentiment.

In the prosecution of this subject, I shall, in the first place, consider what [11] kind of genius, understanding, and temper naturally fit a man for being a physician. – In the second place, what are the moral qualities to be expected from him in the exercise of his profession, viz., the obligation to humanity, patience, attention, discretion, secrecy, and honour, which he lies under to his patients. – In the third place, I shall take notice of the decorums and attentions peculiarly incumbent on him as a physician, and which tend most effectually to support the dignity of the profession; as likewise the general propriety of his manners, his behaviour to his patients, to his brethren, to surgeons, and to apothecaries. – In the fourth place, I shall particularly describe that course of education which is necessary for qualifying a physician to practise with success and reputation; and shall, at the same time, mention those ornamental qualifications expected from the physician as a gentleman of a liberal education, and without which it is difficult to support the honour and rank of the profession. [12]

I begin with an enquiry into the genius, understanding, and temper, which naturally fit a man for being a physician.

Perhaps no profession requires so comprehensive a mind as medicine. In the other learned professions, considered as sciences, there is a certain established standard, certain fixed laws and statutes, to which every question must constantly refer, and by which it must be determined. A knowledge of this established authority may be attained by assiduous application and a good memory. There is little room left for the display of genius, where invention cannot add, nor judgment improve; because the established laws, whether right or wrong, must be submitted to. The only exercise for ingenuity, is in cases where it does not clearly appear what the laws are. But even then, as disputable points must be referred to the determination of judges, whose opinions, being formed from va-[13]rious circumstantial combinations, frequently differ, there is no criterion by which the ingenious reasoner can be judged; and his conclusions, whether well or ill drawn, must still remain undecided. The case is very different in medicine. There we have no established authority to which we can refer in doubtful cases. Every physician must rest on his own judgment, which appeals for its rectitude to nature and experience alone. Among the infinite variety of facts and theories with which his memory has been filled in the course of a liberal education, it is his business to make a judicious separation between those founded in nature and experience, and those which owe their birth to ignorance, fraud, or the capricious systems of a heated and deluded imagination. He will likewise find it necessary to distinguish between important facts, and such as, though they may be founded in truth, are notwithstanding trivial, or utterly useless to the main ends of his profession. Supposing all [14] these difficulties surmounted, he will find it no easy matter to apply his knowledge to practice. In teaching a system of the practice of physic, every disease must be considered separately, and as existing by itself; but in fact diseases are found complicated in endless varieties, which no system, has hitherto been able to comprehend. This occasions an embarrassment to a young practitioner, which nothing can remove but a habit of nice discernment, a quickness of apprehension which enables him to perceive real analogies, and, what is rarely united with this, a solidity of judgment, which secures him from being deceived by imaginary ones. A student of much fancy and some learning, has no idea of this difficulty. In the pride of his heart, he fancies every disease must fly before him; he thinks he not only knows the proximate causes and indications of cure in all distempers, but a variety of remedies that will exactly answer them. It will be unfortunate however for his [15] patients, if a little experience do not humble his pride, and satisfy him that in many cases he neither knows the proximate causes nor the indications of cure, nor how to fulfil these indications when he does know them; or shew him, what is equally humiliating, that the indications are different and contradictory. In this situation his boasted science must stoop, perhaps, for some time, to be an idle spectator, or to palliate the violence of particular symptoms, or to proceed with the utmost fear and diffidence, with such lights as he can receive from a precarious analogy. Such are the difficulties which a physician has to encounter in his early practice; to conquer which is required, besides the qualifications of a proper education, the concurrence of a penetrating genius, and of a clear, solid judgment; and, in many cases, of a quickness of apprehension, instantaneously to perceive where the greatest probability of success lies, and ['resolution' added in errata] to act accordingly. [16]

But, although a physician should possess that enlarged medical genius, which I have just now described, yet talents of another kind are also requisite. A physician has not only for an object, the improvement of his own mind, but he must study the temper, and struggle with the prejudices of his patient, of the relations, and of the world in general; nay, he must guard himself against the ill offices of those, whose interest interferes with his; and it unfortunately happens, that the only judges of his medical merit, are those who have sinister views in concealing or depreciating it. Hence appears the necessity of a physician's having a large share of good sense, and knowledge of the world, as well as a medical genius and learning.

Such are the genius and talents required in a physician; but a certain command of the temper and passions, either natural or acquired, must be added, in order to give them their full [17] advantage. Sudden emergencies occur in practice, and diseases often take unexpected turns, which are apt to flutter the spirits of a man of lively parts and of a warm temper. Accidents of this kind may affect his judgment in such a manner as to unfit him for discerning what is proper to be done, or, if he do perceive it, may, nevertheless, render him irresolute. Yet such occasions call for the quickest discernment and the steadiest and most resolute conduct; and the more, as the sick so readily take the alarm, when they discover any diffidence in their physician. The weaknesses too and bad behaviour of patients, and a number of little difficulties and contradictions which every physician must encounter in his practice, are apt to ruffle his temper, and consequently to cloud his judgment, and make him forget propriety and decency of behaviour. Hence appears the advantage of a physician's possessing presence of mind, composure, steadiness, and an ap-[18]pearance of resolution, even in cases where, in his own judgment, he is fully sensible of the difficulty.

I come now to mention the moral qualities peculiarly required in the character of a physician. The chief of these is humanity; that sensibility of heart which makes us feel for the distresses of our fellow-creatures, and which, of consequence, incites us in the most powerful manner to relieve them. Sympathy produces an anxious attention to a thousand little circumstances that may tend to relieve the patient; an attention which money can never purchase: hence the inexpressible comfort of having a friend for a physician. Sympathy naturally engages the affection and confidence of a patient, which, in many cases, is of the utmost consequence to his recovery. If the physician possesses gentleness of manners, and a compassionate heart, and what Shakespeare so emphatically calls "the milk of human kindness," [19] the patient feels his approach like that of a guardian angel ministering to his relief: while every visit of a physician who is unfeeling, and rough in his manners, makes his heart sink within him, as at the presence of one, who comes to pronounce his doom. Men of the most compassionate tempers, by being daily conversant with scenes of distress, acquire in process of time that composure and firmness of mind so necessary in the practice of physick. They can feel whatever is amiable in pity, without suffering it to enervate or unman them. Such physicians as are callous to sentiments of humanity, treat this sympathy with ridicule, and represent it either as hypocrisy, or as the indication of a feeble mind. That sympathy is often affected, I am afraid is true. But this affectation may be easily seen through. Real sympathy is never ostentatious; on the contrary, it rather strives to conceal itself. But, what most effectually detects this hypocrisy, is a physician's different man-[20]ner of behaving to people in high and people in low life; to those who reward him handsomely, and those who have not the means to do it. A generous and elevated mind is even more shy in expressing sympathy with those of high rank, than with those in humbler life; being jealous of the unworthy construction so usually annexed to it. – The insinuation that a compassionate and feeling heart is commonly accompanied with a weak understanding and a feeble mind, is malignant and false. Experience demonstrates, that a gentle and humane temper, so far from being inconsistent with vigour of mind, is its usual attendant; and that rough and blustering manners generally accompany a weak understanding and a mean soul, and are indeed frequently affected by men void of magnanimity and personal courage, ['in order' added in errata] to conceal their natural defects.

There is a species of good humour different from the sympathy I have been [21] speaking of, which is likewise amiable in a physician. It consists in a certain gentleness and flexibility, which makes him suffer with patience and even apparent chearfulness, the many contradictions and disappointments he is subjected to in his practice. If he be rigid and too minute in his directions about regimen, he may be assured they will not be strictly followed; and if he be severe in his manners, the deviations from his rules will as certainly be concealed from him. The consequence is, that he is kept in ignorance of the true state of his patient; he ascribes to the consequences of the disease, what is merely owing to irregularities in diet, and attributes effects to medicines which were perhaps never taken. The errors which in this way he may be led into, are sufficiently obvious; and might easily be prevented by a prudent relaxation of rules that could not well be obeyed. The government of a physician over his patient should undoubtedly be great. But an [22] absolute government very few patients will submit to. A prudent physician should, therefore, prescribe such laws, as, though not the best, are yet the best that will be observed; of different evils he should chuse the least, and, at no rate, lose the confidence of his patient, so as to be deceived by him as to his true situation. This indulgence, however, which I am pleading for, must be managed with judgment and discretion; as it is very necessary that a physician should support a proper dignity and authority with his patients, for their sakes as well as his own.

There is a numerous class of patients who put a physician's good-nature and patience to a severe trial; those I mean who suffer under nervous ailments. Although the fears of these patients are generally groundless, yet their sufferings are real; and the disease is as much seated in the constitution as a rheumatism or a dropsy. To treat their [23] complaints with ridicule or neglect, from supposing them the effect of a crazy imagination, is equally cruel and absurd. They generally arise from, or are attended with bodily disorders, obvious enough; but supposing them otherwise, still it is the physician's duty to do every thing in his power for the relief of the distressed. Disorders of the imagination may be as properly the object of a physician's attention as those of the body; and surely they are, frequently, of all distresses the greatest, and demand the most tender sympathy; but it requires address and good sense in a physician to manage them properly. If he seems to treat them slightly, or with unseasonable mirth, the patient is hurt beyond measure; if he be too anxiously attentive to every little circumstance, he feeds the disease. For the patient's sake, therefore, as well as his own, he must endeavour to strike the medium between negligence and ridicule on the one hand, and too much solicitude about every trifling [24] symptom on the other. He may sometimes divert the mind, without seeming to intend it, from its present sufferings, and from its melancholy prospects of the future, by insensibly introducing subjects that are amusing or interesting; and sometimes he may successfully employ a delicate and good-natured pleasantry.

We sometimes see a remarkable difference between the behaviour of a physician at his first setting out, and afterwards, when he is fully established in reputation and practice. In the beginning he is affable, polite, humane, and assiduously attentive to his patients: but afterwards, when he has reaped the fruits of such a behaviour, and finds himself independent, he assumes a very different tone; he becomes haughty, rapacious, careless, and often somewhat brutal in his manners. Conscious of the ascendency he has acquired, he acts a despotic part, and takes a most ungenerous advantage of the confidence which people have in his abilities. [25]

A physician, by the nature of his profession, has many opportunities of knowing the private characters and concerns of the families in which he is employed. Besides what he may learn from his own observation, he is often admitted to the confidence of those, who perhaps think they owe their life to his care. He sees people in the most disadvantageous circumstances, very different from those in which the world views them; – oppressed with pain, sickness, and low spirits. In these humiliating situations, instead of wonted chearfulness, evenness of temper, and vigour of mind, he meets with peevishness, impatience, and timidity. Hence appears how much the characters of individuals, and the credit of families, may sometimes depend on the discretion, secrecy, and honour of a physician. Secrecy is particularly requisite where women are concerned. Independent of the peculiar tenderness with which a woman's character should be treated, there are certain circumstances [26] of health, which, though in no respect connected with her reputation, every woman, from the natural delicacy of her sex, is anxious to conceal; and, in some cases, the concealment of these circumstances may be of consequence to her health, her interest, and to her happiness.

Temperance and sobriety are virtues peculiarly required in a physician. In the course of an extensive practice, difficult cases frequently occur, which demand the most vigorous exertion of memory and judgment. I have heard it said of some eminent physicians, that they prescribed as justly when intoxicated as when sober. If there was any truth in this report, it contained a severe reflection against their abilities in their profession. It shewed they practised by rote, or prescribed for some of the more obvious symptoms, without attending to those nice peculiar circumstances, a knowledge of which consti-[27]tutes the great difference between a physician who has genius and one who has none. Intoxication implies a defect in the memory and judgment; it implies confusion of ideas, perplexity and unsteadiness; and must therefore unfit a man for every business that requires the lively and vigorous use of his understanding.

I may reckon among the moral duties incumbent on a physician, that candor, which makes him open to conviction, and ready to acknowledge and rectify his mistakes. An obstinate adherence to an unsuccessful method of treating a disease, must be owing to a high degree of self-conceit, and a belief of the infallibility of a system. This error is the more difficult to cure, as it generally proceeds from ignorance. True knowledge and clear discernment may lead one into the extreme of diffidence and humility; but are inconsistent with self-conceit. It sometimes happens too, [28] that this obstinacy proceeds from a defect in the heart. Such physicians see that they are wrong; but are too proud to acknowledge their error, especially if it be pointed out to them by one of the profession. To this species of pride, a pride incompatible with true dignity and elevation of mind, have the lives of thousand been sacrificed. [29; 30 is blank.]

LECTURE II

Decorums and attentions peculiar to a physician. – How the obligations to these arise. – Duty of a physician with regard to adopting new remedies. – Duty in acquainting a patient and his relations, of his situation. – Conduct of a physician when he despairs of the patient's life. – Conduct in regard to the profits of his profession. – Consultations. – Patients interest not to suffer by the quarrels of physicians. – Behaviour of young physicians to their seniors. – Distinction between physick, surgery, and pharmacy. – Dress. – Manners. – Affectation of delicacy. – Servility. – Remarks on secret medicines. – Charge of infidelity against physicians considered.

I PROCEED now to make some observations on the decorums and attentions peculiar to a physician, and such [31] as tend most effectually to support the dignity of the profession.

Decorum, decency, and propriety, are words very indeterminate in their application; for this reason, that the ideas annexed to them are partly founded in nature and common sense, partly in caprice, fashion, and the customs of particular nations. In the first case, the obligation to them is immutable, the same in all ages and nations; in the latter, it is fluctuating and less binding. When it is necessary, I shall endeavour to mark this distinction.

I have already taken notice of the principal duties a physician owes to his patients, of the propriety of his attending to their tempers and constitutions, and allowing them every indulgence consistent with their safety. Sometimes a patient himself, sometimes one of his friends, will propose to the physician a remedy, which, they believe, may do him service. Their proposal may be a good [32] one; it may even suggest to the ablest physician, what, perhaps, till then, might not have occurred to him. It is undoubtedly, therefore, his duty to adopt it. Yet there are some of the faculty, who, from a pretended regard to the dignity of the profession, but in reality from mean and selfish views, refuse to apply any remedy proposed in this manner, without regard to its merit. But this behaviour can never be vindicated. Every man has a right to speak where his life or his health is concerned, and every man may suggest what he thinks may tend to save the life of his friend. It becomes them to interpose with politeness, and a deference to the judgment of the physician; it becomes him to hear what they have to say with attention, and to examine it with candour; If he really approves, he should frankly own it, and act accordingly; if he disapproves, he should declare his disapprobation in such a manner, as shews it proceeds from conviction, and not from pique or obstinacy. If a pa-[33]tient is determined to try an improper or dangerous medicine, a physician should refuse his sanction, but he has no right to complain of his advice not being followed.

A physician is often at a loss in speaking to his patients of their real situation when it is dangerous. A deviation from truth is sometimes in this case both justifiable and necessary. It often happens that a person is extremely ill; but yet may recover, if he be not informed of his danger. It sometimes happens, on the other hand, that a man is seized with a dangerous illness, who has made no settlement of his affairs, and yet perhaps the future happiness of his family may depend on his making such a settlement. In this and other similar cases, it may be proper for a physician, in the most prudent and gentle manner, to give a hint to the patient of his real danger, and even solicit him to set about his necessary duty. But, in every case, it behooves a [34] physician never to conceal the real situation of the patient from the relations. Indeed justice demands this; as it gives them an opportunity of calling for further assistance, if they should think it necessary. To a man of a compassionate and feeling heart, this is one of the most disagreeable duties in the profession: but it is indispensible. The manner of doing it, requires equal prudence and humanity. What should reconcile him the more easily to this painful office, is the reflection that, if the patient should recover, it will prove a joyful disappointment to his friends; and, if he die, it makes the shock more gentle. Let me here exhort you against the custom of some physicians, who leave their patients when their life is despaired of, and when it is no longer decent to put them to farther expence. It is as much the business of a physician to alleviate pain, and to smooth the avenues of death, when unavoidable, as to cure diseases. Even in cases where his skill as a physician can be of no further avail, his [35] presence and assistance as a friend may be agreeable and useful, both to the patient and to his nearest relations. Neither is it proper that he should withdraw when a clergyman is called to assist the patient in his spiritual concerns. On the contrary, it is decent and fit that they should mutually understand one another and act together. The conversation of a clergyman, of cheerful piety and good sense, in whom a sick man confides, may sometimes be of much more consequence in composing the anguish of his mind, and the agitation of his spirits, than any medicine; but a gloomy and indiscreet enthusiast may do great hurt; may terrify the patient, and contribute to shorten a life that might otherwise be saved.

There are often unhappy jealousies and animosities among those of the profession, by which their patients may suffer. A physician, however, who has any sense of justice or humanity, will never involve his patient in the consequences of private [36] quarrels, in which he has no concern. Physicians in consultation, whatever may be their private resentments or opinions of one another, should divest themselves of all partialities, and think of nothing but what will most effectually contribute to the relief of those under their care. If a physician cannot lay his hand to his heart, and say that his mind is perfectly open to conviction, from whatever quarter it shall come, he should in honour decline the consultation. Many advantages arise from two physicians consulting together, who are men of candour, and have mutual confidence in each other's honour. A remedy may occur to one which did not to another; and a physician may want resolution, or sufficient confidence in his own opinion, to prescribe a powerful but precarious remedy, on which, however, the life of his patient may depend; in this case the concurring opinion of his brother may fix his own. But, if there is no mutual confidence; if opinions be regarded, [37] not according to their intrinsic merit, but according to the person from whom they proceed; or, if there be reason to believe, that sentiments delivered with openness are to be whispered abroad, and misrepresented to the publick, without regard to the obligations of honour and secrecy; and if, in consequence of this, a physician is singly to be made responsible for the effects of his advice; in such cases, consultations of physicians tend rather to the detriment than to the advantage of the sick: and the usual and indeed most favourable conclusion of them is some very harmless but insignificant prescription.

The quarrels of physicians, when they end in appeals to the public, generally hurt the contending parties; but, what is of more consequence, they discredit the profession, and expose the faculty itself to ridicule and contempt.– Nothing, in my opinion, but [substitute 'the following' for 'this' in errata] this cause, can justify any physician for refusing to [38] consult with another, when he is required to do so. If he be conscious he cannot behave with temper, and that his passions are so ruffled as to impair his judgment, he may and ought to refuse it. But such circumstances, as the university where the person he is to consult with had his degree, or indeed whether he had a degree from any university or not, cannot justify his refusal. It is a physician's duty to do every thing in his power that is not criminal, to save the life of his patient, and to search for remedies from every source, and from every hand, however mean and contemptible. This, it may be said, is sacrificing the dignity and interests of the faculty. But, I am not here speaking of the private police of a corporation, or the little arts of craft. I am treating of the duties of a liberal profession, whose object is the life and health of the human species, a profession to be exercised by gentlemen of honour and ingenuous manners; the dignity of which can [39] never be supported by means that are inconsistent with its ultimate object, and that only tend to increase the pride and fill the pockets of a few individuals.

It becomes young physicians to be particularly attentive to the propriety of their behaviour when consulting with their seniors. Besides the respect due to age, these are entitled to a particular deference from their longer and more extensive experience. The revolutions indeed of medical hypotheses and systems are so quick, that an old and a young physician seldom reason in the same way on subjects of their profession; although the difference be sometimes rather apparent than real, when they use only a different language to express sentiments essentially the same. But it generally happens, that the speculations which principally engage the attention of young physicians, seldom in any degree affect their practice; and therefore, as they are in a great measure foreign to the [40] business, they should never introduce them in medical consultations. They shew equal want of sense and good-manners, when they wantonly take opportunities of expressing a contempt for opinions as antiquated and exploded, in which their seniors have been educated, and which they hold as firmly established. A little reflection might teach them, that it is not impossible, but in the course of a few years, their own most favourite theories may be discovered to be as weak and delusive as those which have gone before them; and this should lead them to consider how sensibly they may be hurt themselves, when they find those idols of their youth attacked by the petulant ridicule of the next generation; when, perhaps, they are arrived at a time of life in which they have neither patience [substitute 'abilities' for 'patience' in errata] nor temper to defend them.

The same respect that ought to be shewn to the opinions of elder physicians, should be extended to their favourite [41] authors, and indeed to all such writers in medicine as have contributed to its advancement, and whose names, in the successive ages of physic have been revered by the wisest and most learned of the profession. It is equally prudent and decent for young practitioners, when they differ in opinion from the rest of mankind, to express their dissent with modest and good-manners. Their abuse of characters which have been generally esteemed, has more the appearance of petulance and self-conceit, than of the liberal and ingenuous spirit that flows from the love of truth. There is, indeed, an ardent love of freedom, and an impatience of the controul of authority in all matters of sentiment and reasoning, which is both natural and proper in young men. This high spirit is very properly shewn when they expose to ridicule such authors as are remarkably vain and ostentatious, or when they chastise the insolence of such as are assuming, supercilious, and dictatorial; but, in speaking of such men as [42] Hippocrates, Sydenham, or Boerhaave, who were no less eminent for their candour and modesty, than for their genius and merit in their profession, at the same time that their opinions are canvassed with freedom, their characters should be treated not only with decency but with reverence.

There have arisen at different periods, and particularly in France, about twenty years ago, great disputes about the boundary of physic and surgery, and the proper subordination of surgery to medicine. A dispute hurtful to mankind, and which has been often conducted in a manner unworthy of scholars and gentlemen. I shall embrace this opportunity of giving my sentiments concerning it.

There was anciently, as Celsus informs us, a division of medicine into three parts: the first regarded the regulation of diet; the second, the prescription of remedies; the third, manual operations, [43] or surgery. The two first, though distinguished in theory, were always united in practice; the last has often been exercised separately. Sometimes the ancient physicians performed the manual part themselves; at other times, it was done by slaves kept for that purpose. Among the moderns, the arts of physic and surgery have often been promiscuously practised by the same persons; for example, Hildanus[14], Severinus[15], Bartholine[16], and many others of distinguished genius and learning. But, in many parts of Europe, both now and formerly, surgery has not been reckoned among the liberal professions, but surgeons have ignominiously been classed with the corporation of barbers. In such places, we may reasonably suppose, that this art must be often practised by people of the lowest rank, who have never received a liberal education. The separation of physic from surgery in modern times, has been productive of the worst consequences. The physicians and surgeons, [44] formed into separate societies, had separate interests to support, which, in many cases, clashed with each other. The surgeons claimed not only the exclusive privilege of performing all operations, but likewise the management of most external diseases, and some internal ones, where operations were supposed to be often necessary; by which means the method of cure in many diseases was sometimes left to the direction of ignorant as well as illiterate men. But it must be apparent to every sensible and ingenuous observer, that the diseases of the human body are so intimately connected, that it is impossible to understand some of them perfectly, and be entirely ignorant of all the rest;

and hardly possible to understand any of them, without some knowledge of Anatomy, and of the Animal oeconomy, both in its sound and morbid state. It must at the same time be owned, that a practitioner, well-grounded in such general knowledge, may have considerable advantages; and more rea-[45]dily make improvements, by attaching himself to the study of one or two particular diseases. – Every distemper, external as well as internal, falls under the cognizance of the physician, and it is a reflection on him to be ignorant of any of them; neither is it possible to fix any such precise boundaries between external and internal diseases, as to render the distinction in any degree useful, or applicable in practice. Suppose a person to break his leg, and a fever and gangrene to ensue; the question occurs, whether the limb should be immediately amputated, or to ['whether we should' substituted for 'to' in errata] wait for some time till the effects of certain medicines, given with a view to stop the progress of the mortification, are known. It is evidently the business of a physician, in this case, to judge from the symptoms, from the habit of body, and from other circumstances, whether the delay is prudent or not. – As to the performance of the operation itself, that is a different question. The genius and education requisite to [46] make a good physician, are not necessary to make a good operator. – What is peculiarly necessary to make a good operator, is a resolute, collected mind, a good eye, and a steady hand. These talents may be united with those of an able physician; but they may also be separated from them. – If surgery were confined to a set of men who were to be merely operators, it might justly be expected, that the art would be more quickly brought to perfection by such men, than by those who follow a more complicated business, and practise all the branches of medicine. The same advantage would accrue to pharmacy, if apothecaries were to be confined to the mere business of preparing medicines. But, in reality, this is not the case. In some parts of Europe surgeons act as physicians in ordinary; in others, the apothecaries do this duty, without a medical education. The consequence is, that in many places physic is practised by low, illiterate men, who are a disgrace [47] to the profession. In regard to pharmacy, it were much to be wished, that those who make it their business should have no connexion with the practice of physic, or that physicians should dispense their own medicines, and either not charge the expense of them to their patients at all, or ['charge it' added in errata] at the prime cost. It is only in one or other of these ways, that we can ever hope to see that simplicity of prescription take place in the practice of medicine, which all who understand its real interests so ardently wish for; and it is only from such an arrangement that we can expect to see physicians placed in that honourable independence, which subjects them to no attentions but such as tend to the advancement of their art. But it is a known fact, that, in many parts of Europe, physicians who have the best parts, and best education, must frequently depend for their success upon apothecaries, who have no pretensions either to the one of the other; and that the obligation to apo-[48]thecaries is too often repaid by what every one concerned for the honour of medicine must reflect on with indignation.

From what I have said, it is evident, that I have no intention to throw reflections upon any particular branch of the profession. Every department of it is respectable, when exercised with capacity and integrity. I only contend for an evident truth, either that the different branches should be separately professed, or, if one person will profess all, he should be regularly educated to, and thoroughly master of all. I am not here adjusting points of precedence, or insinuating the deference due to degrees in medicine. As a doctor's degree can never confer sense, the title alone can never command regard; neither should the want of it deprive any man of the esteem and deference due to real merit. If a surgeon or apothecary has had the education, and acquired the knowledge of a physician, he is a phy-[49]sician to all intents and purposes, whether he has a degree or not, and ought to be respected and treated accordingly. In Great Britain, surgery is a liberal profession. In many parts of it, surgeons or apothecaries are the physicians in ordinary to most families, for which trust they are often well qualified by their education and knowledge; and a physician is only called where a case is difficult, or attended with danger. There are certain limits, however, between the two professions, which ought to be attended to; as they are established by the customs of the country, and by the rules of their several societies. But a physician, of a candid and liberal spirit, will never take advantage of what a nominal distinction, and certain privileges, give him over men, who, in point of real merit, are his equals; and will feel no superiority, but what arises from superior learning, superior abilities, and more liberal manners. He will despise those distinctions founded in vanity, self-interest, or caprice; and [50] will be careful, that the interests of science and of mankind shall never be hurt, on his part, by a punctilious adherence to formalities.

Among the peculiar decorums of a physician's character, much stress has been laid on a certain formality in dress, and a particular gravity in his behaviour. I have before observed, that decorum and propriety have their foundation sometimes in nature and common sense, sometimes only in caprice and fashion. This observation may be exemplified by the present subject. In many stations a particular formality and pomp of dress is highly proper, independently of any fashion whatever; for example, in judges and in magistrates. Whatever circumstances in their mode of dress, or external appearance, make them the objects of awe and reverence, are necessary and decent; because they impress the minds of the people with a due veneration, and fear of the laws. Neither is there any danger of abuse from this reverence procured to [51] the office of a magistrate. The case is very different in the profession of medicine. There is no natural propriety in a physician's wearing one dress in preference to another; it not being necessary that any particular respect or authority should be annexed to his office, independent of what his personal merit commands. Experience, indeed, has shewn, that all our external formalities have been often used as snares to impose on the weakness and credulity of mankind; that in general they have been most scrupulously adhered to by the most ignorant and forward of the profession; that they frequently supplant real worth and genius; and that, far from supporting the dignity of the profession, they often expose it to ridicule. If then there be no natural and real propriety in a physician's wearing a distinguishing dress, he can be under no obligation to use it, but what arises from the particular fashion of the country where he resides. This is an obligation, however, which common sense and pru-[52]dence make it necessary he should regard. If the customs or prejudices of any country affix the idea of sense, knowledge, or dignity to any mode of dress, it is a physician's business, from motives of prudence, to equip himself accordingly. But, in a country, where a physician's capacity is not measured by such standards, and where he may dress like other people, without sinking in their estimation, I think he is at full liberty to avail himself of this indulgence, if he so chuse, without being considered as deviating from the propriety and decency of his profession.

In some cases, there is a great impropriety in a physician's having any distinguishing formality in his dress or manners. I do not hint merely at the disagreeable impression, not to say terror, with which this sometimes affects the minds of children. Even among people who possess the greatest vigour and firmness of mind when in health, there is often a feebleness [53] and depression of spirits attendant on sickness, that renders the sight of any stranger whatever very painful. In such a state of mind, the visit of a physician, even when wished for, is often particularly dreaded, as it naturally awakens the apprehensions of danger; apprehensions, which a formal dress, and a solemn behaviour, are ill calculated to dispel. Surely, if there be, at any time, a propriety in an easy, cheerful, soothing behaviour, it must be on such an occasion, where it is so necessary to forget the physician in the friend.

I see, indeed, no reason why the general character of a physician's manners should be any way singular. They may be affable without meanness, grave without formality, and cheerful without levity. They will naturally vary according to the circumstances in which he is placed. How different the appearance of a physician rejoicing with his patient at his restoration to health and spirits, or communi-[54]cating to his friends the accounts of his approaching dissolution! If, however, the manners of a country require that he should observe the same unvaried face and solemnity, unmoved by every object around him, and equally susceptible of joy or sorrow, he must submit; but if, without such necessity, he voluntarily lays himself under these or any other restraints, and assumes a fictitious character, there is reason to suspect the qualities of his heart or of his understanding.

A physician should carefully guard against any little peculiarities stealing into his manners, which can in any degree render him an object of ridicule. Young physicians, in particular, will much deceive themselves, if they imagine they can indulge in such particularities with the same impunity that their seniors sometimes do. It is indeed an observation, which perhaps does no great honour to mankind, that when once a physician's reputation for knowledge in his profession [55] comes to be thoroughly established, almost every peculiarity of manners, even some that would be in other men offensive, deepen the impression made on the imagination by his supposed merit, and increase his popularity and fame.

There is great impropriety in a physician's indulging himself in a certain delicacy, which makes him liable to be disgusted with some disagreeable circumstances he must unavoidably meet with in his practice. Genuine delicacy is a virtue of the mind, and though it shews itself by an attachment to cleanliness, neatness, and even elegance where it can be afforded, yet it always gives place and forgets itself, where duty or the interests of humanity require it. It is a mistake in a physician to think any attentions, or duties, below his dignity, which can contribute to the relief of his patient. When necessity calls, he acts unworthily, if he do not become, to the best of his abilities, surgeon, apothecary, and even [56] nurse. If, however, without such necessity, he encroach on another's province, then, indeed, he degrades himself; not because he acts below the dignity of a physician, but because he behaves in a manner unbecoming the character of a gentleman.

The attendance given to a patient should be in proportion to the urgency and danger of his complaints. As the physician is the best judge of this, he should regulate his visits accordingly. But some delicacy is often required, to prevent such frequent visits as may be necessary, from bringing an additional expense upon the patient. A patient is entitled to the whole attention of his physician, while he remains with him. Whatever other business or avocations he may have, he should dedicate that time entirely to him. That continual hurry which some of our profession seem to be in, is sometimes mere affectation; but it often proceeds from other causes. Some keep themselves con-[57]stantly embarrassed by a want of œconomy of their time, and of a proper arrangement of their business; some, from a liveliness of imagination, and an unremitting activity of mind, involve themselves in such a multiplicity of pursuits as cannot be overtaken. But from whatever source it arises, it ought to be timely corrected, and not suffered to go into a habit. It prevents a physician from doing his duty to the sick, and at the same time weakens their confidence in him.

Having freely expressed my sentiments concerning what I think wrong in the conduct of some of our faculty, I shall now, with the same freedom, animadvert on a particular circumstance not unfrequent in the behaviour of other learned men, as well as physicians, which seems to me essentially injurious to the dignity of our profession; I mean that servility of manners towards people of rank and fortune, which so often disgraces men, in other respects eminent for learning [58] and ingenuity. The external magnificence and splendor attendant on high rank is apt sometimes to dazzle their understanding, and makes them pay too much veneration to those outward distinctions of title and fortune, which their philosophy ought to make them [d]espise. ['d' is missing in original]

Great disputes have arisen in our profession, about the propriety of a physician's keeping secrets or nostrums. It has been said, with some plausibility, in vindication of this practice, that the bulk of mankind seldom attend, or pay much regard, to what is made level to their capacities; and that they are apt to undervalue what costs them nothing. Experience shews, that men are naturally attached to whatever has an air of mystery and concealment. A vender of a quack medicine does not tell more lies about its extraordinary virtues, than many people do who have no interest in the matter; even men of sense and pro-[59]bity. A passion for what is new and marvellous, operates more or less on the imagination; and, in proportion as that is heated, the understanding is perplexed. When a nostrum is once divulged, its wonderful qualities immediately vanish, and in a few months it is generally forgotten. If it be really valuable, the faculty perhaps adopt it: but it never recovers its former reputation.– It is likewise said, that this is the only way in which any good medicine can be introduced into practice; as the bulk of mankind will more readily follow the directions of a man who professes to cure them by mysterious means, than those of a regular physician, who prescribes plain and common remedies. It is further alleged, that some of the best medicines were originally introduced as secrets, though opposed by the regular physicians. But allowing this to be true, yet I am persuaded, that nostrums, on the whole, do more harm than good; that they hinder [60] the advancement of the art, by making people neglect what is known and approved, in pursuit of what is unknown, and probably never to be divulged; and that, from their being generally dispensed by low and illiterate men, who prescribe them indiscriminately, they are become a public nuisance in these kingdoms. – In some places on the continent, however, physicians of honour and reputation keep nostrums. In such hands, the same abuses will not be committed, as we experience here; but still the practice has an interested and illiberal appearance.

Curiosity in a patient or his friends to know the nature of the medicines prescribed for him is natural, and therefore not blameable; yet this is a curiosity which it is often very improper to gratify. There is a natural propriety in mankind to admire what is covered with the veil of obscurity, and to undervalue whatever is fully and clearly explained [61] to them. A firm belief in the effects of medicine depends more on the imagination, than on a rational conviction impressed on the understanding; and the imagination is never warmed by any object which is distinctly perceived, nor by any truth obvious to common sense. Few people can be persuaded that a poultice of bread and milk is in many cases as efficacious as one compounded of half a dozen ingredients, to whose names they are strangers; or that a glass of wine is, in most cases where a cordial is wanted, one of the best that can be administered. This want of faith in the effects of simple known remedies, must of necessity occasion a disregard to the prescription, as well as create a low opinion of the physician. Besides, where a patient is made acquainted with the nature of every medicine that is ordered for him, the physician is interrupted in his proceedings by many frivolous difficulties, not to be removed to the satisfaction of one ignorant of medicine. The consequence [62] of this may be to embarrass the physician, and render him irresolute in his practice; particularly in the administration of the more powerful remedies. It should be further considered, that when a patient dies, or grows worse under the care of a physician, his friends often torment themselves, by tracing back all that has been done, if they have been made acquainted with it, and may thus be led, very unjustly, to charge the physician with what was the inevitable consequence of the disease. There are, indeed, cases where it may be proper to acquaint a patient with the nature of the remedies, as there sometimes peculiarities in a constitution in regard both to the quality and quantity of the medicine, which a physician ought to be informed of before he prescribes it.

I shall conclude this subject with some observations on a charge of a heinous nature, which has been often urged against our profession; I mean that of [63] infidelity, and contempt of religion. I think the charge ill-founded; and will venture to say, that the most eminent of our faculty have been distinguished for real piety. I shall only mention, as examples, Harvey[11], Sydenham[1], Arbuthnot[17], Boerhaave[18], Stahl[5], and Hoffman[19]. – It is easy, however, to see whence this calumny has arisen. Men whose minds have been enlarged by knowledge, who have been accustomed to think, and to reason upon all subjects with a generous freedom, are not apt to become bigots to any particular sect or system. They can be steady to their own principles without thinking ill of those who differ from them; but they are impatient of the authority and controul of men, who would lord it over their consciences, and dictate to them what they are to believe. This freedom of spirit, this moderation and charity for those of different sentiments, have frequently been ascribed, by narrow-minded people, to secret infidelity, scepticism, or, at least, to lukewarmness in [64] religion; while some who were sincere Christians, exasperated by such reproaches, have sometimes expressed themselves unguardedly, and thereby afforded their enemies a handle to calumniate them. This, I imagine, has been the real source of that charge of infidelity so often and so unjustly brought against physicians. In a neighboring nation, where few people have been used to think or to reason with freedom on religion, and where, till of late, no man durst express himself freely on the subject, some ingenious and lively writers have, within these few years appeared, who, impatient to display their newly-acquired liberty, have attempted to shake the foundations of all religion, natural as well as revealed. Lately emancipated from the lowest superstition, by a transition not unnatural, they have plunged at once into Atheism. It is perhaps for the better, that these gentlemen have carried matters so far; because [65] it is to be hoped the evil will soon cure itself. Mankind may have their religious opinions disfigured by various superstitions: but still religion is natural to the human mind, and every attempt to eradicate it, will be found as impotent as it is wicked. But, supposing that Atheism came universally to prevail, together with the disbelief of the immortality of the soul, the duration of such sentiments would necessarily be very short; because they would at once unhinge all the bonds of society, and produce a continued scene of anarchy and wickedness. Divested of that uncouth, metaphysical dress, under which they long lay concealed, the gloomy speculation of a few recluse men, they are now produced to the world, adorned with what passes among many for wit and humour, and adapted to every capacity. So far as they contain any argument, their weakness has been often demonstrated. One method taken by [66] the present patrons of infidelity to propagate their opinions is somewhat dangerous. With much assurance, they insinuate, that all who avow their belief in natural or revealed religion, are either hypocrites or fools. This is attacking youth on their weak side. A young man, of a liberal spirit, naturally disdains the idea of hypocrisy; and, from an ill-judged pride, is afraid of whatever may subject him to so mean an imputation. Vanity, again, is their most ruling passion, as they commonly dread contempt above everything, and resent reflections on the weakness and narrowness of their understanding, more than any charge against their principles or morals. But I will venture to say, that men of the most enlarged, clear, and solid understandings, who have acted with the greatest spirit, dignity, and propriety, and who have been regarded as the most useful and amiable members of society, have never openly insulted, or insidiously [67] attempted to ridicule the principles of religion; but, on the contrary, have been its best and warmest friends. – The study of medicine, of all others, should be the least suspected of leading to impiety. An intimate acquaintance with the works of Nature, raises the mind to the most sublime conceptions of the Supreme Being, and at the same time dilates the heart with the most pleasing views of Providence. The difficulties that necessarily attend all deep enquiries into a subject so disproportionate to the human faculties, should not be suspected to surprize a physician, who, in his practice, is often involved in perplexity, even in subjects exposed to the examination of his senses.

There are, besides, some peculiar circumstances in the profession of a physician, which should naturally dispose him to look beyond the present scene of things, and engage his heart on the side of religion. He has many opportunities [68] of seeing people, once the gay and the happy, sunk in deep distress; sometimes devoted to a painful and lingering death; and sometimes struggling with the tortures of a distracted mind. Such afflictive scenes, one should imagine, might soften any heart not dead to every feeling of humanity, and make it reverence that religion which alone can support the soul in the most complicated distresses; that religion, which teaches to enjoy life with cheerfulness, and to resign it with dignity. A physician, who has the misfortune to disbelieve in a future state, will, if he have common good-nature, conceal his sentiments from those under his charge, with as much care as he would preserve them from the infection of a mortal disease. With a mind unfeeling, or occupied in various pursuits, he may not be aware of his own unhappy situation; yet it is barbarous to deprive expiring nature of its last support, and to blast the only surviving comfort of those who have taken a last farewell of [69] every sublunary pleasure. But, if motives of humanity, and a regard to the peace and happiness of society cannot restrain a physician from expressing sentiments destructive of religion or morals, it is vain to urge the decency of the profession. The most favourable construction we can put on such conduct, is to suppose that it proceeds from an ungovernable levity, or a criminal vanity, that forgets all the ties of morals, decency, and good manners.

I shall make no apology for going a little out of my way in treating of so serious a subject. In an enquiry into the office and duties of a physician, I deemed it necessary to attempt to wipe off a reflection so derogatory to our profession; and, at the same time, to caution you against that petulance and vanity in conversation, which may occasion imputations of bad principles, equally dangerous to society, and to your own interest and honour. [70]

LECTURE III

Connection of the several branches of physic with its practice. – Impropriety of wasting too much time in the study of these branches. – Necessity of a knowledge of anatomy, and physiology – of natural philosophy, and chemistry. – Laws of union between the soul and body, and of the nervous system, not explicable upon mechanical or chemical principles. –

Comparative anatomy. – Pathology, – Theory of physic, what it properly signifies. – Materia medica.– Botany. – Natural history. – Necessity of a physician's being well founded in these preliminary sciences. – Ornamental qualifications. – Knowledge of the history of physic – of mathematics – of the Latin, Greek, and French languages – of our native language. – Observations on the style and composition proper for medical writings.

I PROCEED NOW to explain the connexions of the several branches of physic [71] with the practical part of it, and enquire how far a previous knowledge of these is necessary, in order to practise with reputation and success.

Here I must previously observe, that all the works of Nature are so intimately connected, that no one part of them can be well understood by considering and studying it separately. In order, therefore, to be qualified for the practice of physic, a variety of branches of knowledge, seemingly little connected, are nevertheless necessary. As this is the case, it is proper that a student should be on his guard not to waste his time and labour in pursuits which either have no tendency, or a remote one, to throw light on his profession. Life is too short for every study that may be deemed ornamental to a physician; it will not even allow time for every study that has some connection with physic. Every one of the sciences I am about to name, is of great extent; but it will be necessary for a physician to confine his appli-[72]cation to such parts of them as are really subservient to practice. If a student's genius incline him more particularly to any of these preliminary sciences, he may, if he pleases, indulge himself in it, taking care not to impose on himself, and consider this as studying physic.

The necessity of a previous knowledge of anatomy to the practice of physic, is apparent, and needs no illustration.

The necessity of the knowledge of physiology, which comprehends the doctrine of the animal fluids, and of all the functions in their sound state, is equally evident. – When you enquire into this subject, you find the human body a machine, constructed upon the most exact mechanical principles. In order, then, to understand its movements, you must be well acquainted with the principles of mechanics. Considering the human body in another view, you find fluids of different kinds circulating through tubes of [73] various diameters; the laws of their motions, therefore, cannot be understood, without a knowledge of hydraulics. The eye appears to be an admirable optical machine; and, of course, the phenomena of vision cannot be explained, without a knowledge of the principles of optics. As the human body is surrounded with an elastic fluid, the air, subject to various changes, in respect of gravity, heat, moisture, and other qualities which have great influence on the constitution, it is proper to be acquainted with the nature and properties of this fluid; the knowledge of which constitutes the science of pneumatics. It were easy to bring many more examples, to shew how necessary a knowledge of the various branches of natural philosophy is to the right understanding of the animal œconomy.

But the different phenomena of the animal œconomy are not all to be explained from common mechanics; va-[74]rious changes are induced upon the fluids from chemical principles. It is, therefore, necessary to be acquainted with the chemical history of our fluids, and with the chemical analysis of whatever is taken into the human body as food or medicine, and, in general, of all the substances which can, in any degree, influence it. This shews the necessity of a knowledge of chemistry, previous to the study of the practice of physic.

Yet the most accurate knowledge of anatomy, and of the principles of mechanics and chemistry, are insufficient to explain all the phenomena in the body. The animal machine differs in many respects from an inanimate one. The former has a power of beginning motion within itself. An internal principle directs and influences most of the operations of the body, by a set of laws totally distinct from, and independant of, any principles of mechanics or chemistry [75] hitherto known. An animal body likewise differs from a common machine, in having a power, to a certain degree, of curing its own disorders, and rectifying many deviations from its natural state; as in the case of fractured bones, incarnation of wounds, enlargement of one kidney when the other is destroyed or rendered useless, and in the successful efforts of Nature in the cure of many diseases. Many feeble attempts have been made to explain the phenomena of the animal body upon mechanical and chemical principles alone; but without success.

The laws of the nervous system, though of the most difficult investigation, are equally steady and regular with any other laws of Nature; so are the laws relating to the mutual influence of the mind and body upon each other; a very important enquiry to a physician: This leads to an extensive and interesting subject, the history of the faculties of the [76] human mind, which, if we are not on our guard, is apt to lead us insensibly into a labyrinth of metaphysics. A student of genius should be watchful, lest his attention be too deeply engaged by this specious kind of philosophy, which gives so much room for imagination, and so little for experiment; apparently ingenious, but really triffling and useless: a philosophy, in short, which, by keeping the mind incessantly employed about subtleties of its own creation, soon renders it incapable of a patient and severe investigation of Nature.

In order to illustrate the human physiology, a knowledge of the comparative anatomy of some animals that most nearly resemble man, is requisite. Several important discoveries in the animal œconomy have been illustrated by experiments first made on brutes, many of which could not have been made on the human subject, e.g. the experiments relating to the circulation of the blood, [77] respiration, muscular motion, sensibility and irritability of different parts of the body, and the effects of various medicines. The instinct of brutes have sometimes given the first hint of valuable remedies, and might throw light on the subject of regimen, and the cure of diseases, if they were properly attended to. At the same time it must be acknowledged, that the comparative anatomy has often led into great mistakes, by too hastily applying it to the human body.

The writers on physiology have usually considered the body as a permanent subject, exhibiting uniformly the same appearances; but, in applying the knowledge of the animal œconomy to practice, it is necessary to consider the human constitution as perpetually fluctuating, and not, perhaps, exactly the same in any two persons. It were endless to trace the diversity of constitutions among mankind, neither would it be an enquiry of much utility; but there are [78] some varieties which it is necessary to attend to. These depend chiefly on the difference of age, sex, climate, and manner of living; and some original temperaments, or habits of body, not produced by any of these circumstances. It belongs to physiology, to enquire into the laws of the union between the mind and the body; into the effects of culture and education upon the constitution; into the power of habit, the effects of enthusiasm, and force of imagination. This short detail shews how extensive a study physiology is, and how intimately connected with the study of the practice of medicine.

As physiology considers the whole appearances of the animal oeconomy in its sound state, pathology considers those in a morbid state. It delivers the general doctrine of the causes, effects, and symptoms of diseases. The *therapeia* treats of the general laws to be observed in the cure of diseases, and of the general na-[79]ture of the remedies. This includes surgery and the *materia medica*. The usefulness of a knowledge of mechanics appears most evidently in the practice of surgery. This art has, in fact, received the greatest improvement within these hundred years, since the doctrine of mechanics came to be more generally understood.

The physiology, pathology, and therapeutics, form what are called the Institutions of Medicine, and, by some, the Theory of Physic. The world has been so long abused by ill-founded, though sometimes plausible hypotheses, assuming the same Theories, that a general prejudice now prevails against the very expression, *Theory of Physic*, as if it contained nothing but useless disquisitions, a display of arrangement, subtle distinctions, chiefly nominal, and an establishment of general principles, many of them false, and others so vague or ambiguous as to be incapable of any [80] useful application. But this is a false representation of the theory or institutions of medicine. These ought to contain every fact tending to illustrate the animal œconomy either in its sound or diseased state, and every fact that may be useful to a physician in forming indications of cure. These facts ought to be accurately arranged, so as either to establish general principles, or at least to point at them, especially at such as lead to practice. In this view the institutions become a natural preliminary to the practice of physic.

A knowledge of the *materia medica* is intimately connected with the practice of medicine. It contains the doctrines of the instruments with which a physician operates, and a history of the effects of medicines. In this branch the use of chemical knowledge is apparent. It teaches how to preserve and separate the useful parts of medicines. But, in order to understand pharmacy, the know-[81]ledge of chemistry is indispensible. For want of this knowledge, at least for want of a proper application of it, true pharmacy has, till of late, been little understood.

The effects of medicines on the human body are sometimes to be explained upon mechanical, sometimes upon chemical, principles; but much oftener depend on the effects they produce upon the nervous system; and, in consequence of these, upon an excess or deficiency of the various animal motions and secretions.

The science of botany is subservient to the practice of physic, so far as it facilitates the knowledge of plants by reducing them into the most commodious system; and, though it is not necessary for a physician to be particularly acquainted with the name and history of every plant, yet he ought to be so well grounded in the principles of botany, as [82] to be able to find its place in the system, and to be particularly acquainted with those plants which are either used in diet or in medicine.

I have now shortly explained the connection of the several branches of physic with the *praxis medica*, which comprehends the *hygiene*, or the method of preserving health and prolonging life, and the application of general pathology, and general therapeutics, to the history and cure of particular diseases. – It will naturally then be asked, Is a person unqualified for the practice of physic and surgery, who is not master of all these parts of learning, which have been deemed to be necessary preliminaries? To this it may be answered, That one may, in some measure, practice medicine as he may do a mechanic art, without much knowledge of its principles. A sailor may navigate a ship, who knows little of the principles of navigation: and a person may make a dial, who [83] knows nothing of astronomy, spherical trigonometry, or the projection of the sphere. It is the same in all the other practical arts of life: and yet, in all these, there are obvious advantages arising from a knowledge of the principles on which they are founded. But in medicine, the necessity of being acquainted with the *principia* of the art is much greater; because there can be no general rules laid down for the practice of physic, which can be applied in all cases. Difference of age, of constitution, of climate, and many other circumstances, occasion variations in the application of the plainest remedies that can be prescribed; and, without a knowledge of the principles of his profession, a physician must be often at a loss. It will be readily acknowledged, that there have been many physicians successful in practice, who, at the same time, were deficient in the knowledge of the foundations of medicine. But this has been owing to their uncommon genius and sagacity, which [84] enabled them to apply what little knowledge they had with judgment, and consequently with success; while, perhaps, another physician, better founded in his profession, for want of this natural genius and sagacity, has blundered in his practice, by a wrong application of his knowledge. Besides, as medicine is so complicated a science, many of those who study it regularly, take a particular attachment to some of its parts; and these so far engage their attention, that they neglect the study of the other branches. In consequence of this, some of our profession have been distinguished for their skill in anatomy, chemistry, and botany, who, nevertheless, have been very indifferent physicians. But, although a profound knowledge of these sciences is insufficient of itself to make a good practical physician; yet no man of sense will infer from thence that they are not eminently useful. [85]

I do not insist here on the necessity of a minute knowledge of these sciences; nor indeed could time be spared to acquire it. A particular acquaintance with the appearance of the muscles, in all the various motions and attitudes of the body, is a study more necessary to a painter, or to a statuary, than to a physician; and, in this view, they ought to be expert anatomists. If chemistry is prosecuted in its full extent and application to all the useful arts, it is a boundless study: so is botany, if we would be acquainted with every circumstance of every plant. It is therefore necessary, that a student, while he endeavours to make himself master of the leading and fundamental principles of these sciences, should always have an eye to their application to his own profession, and bend his attention to that quarter.

On the whole, I hope, it will appear, that a physician who understands the [86] principles of his profession, who has an extensive acquaintance with every branch of natural knowledge connected with it, who properly applies his knowledge, and who has genius and attention equal to others, must have a great advantage, as a practical physician, over one who is ignorant of the principles of medicine, and of the sciences connected with it. Genius and sense are, indeed, the peculiar gifts of Heaven, and cannot be acquired by the most extensive learning, or the greatest efforts of industry. But, with these assistances, genius and sense are capable of the highest improvement.

Besides the above mentioned branches of learning, which are in a manner essential to the rational practice of physic, there are others, which, though they may be deemed rather ornamental, yet an accomplished physician should not be ignorant of. [87]

It may be reasonably expected, that every gentleman should be acquainted with the history of the science which he professes. The history of medicine is not a subject of mere curiosity. To a physician, it is an useful and interesting enquiry. It is indeed an unpleasant task, and, at first view, seems an useless one, to enquire into the numerous theories that have influenced the practice of physic in different ages, Of these there has been a succession, which, in their turn, have been admired, and which have greatly influenced the practice of physic, and afterwards sunk into deserved oblivion. If their bad effects had ceased with the follies which gave them birth, it would have been unnecessary to revive their memory. But this has not been the case. A wrong practice, introduced in consequence of a prevailing theory, soon becomes diffused among people who are no judges whether the theory itself be well or ill founded. A physician of spirit and ingenuity, per-[88]haps, rises up, and shews the absurdity of the theory; but it is not in his power to remove its pernicious consequences in practice. These were soon spread among a thousand ignorant people, who had adapted them to a theory of their own: for it must be observed, that the most illiterate pretenders to physic have their theories; and such pretenders, partly from ignorance, partly from pride, and partly from habit, are, of all others, the most obstinately attached to them.

A thorough knowledge of the history of physic, by discovering the sources of the maxims and remedies adopted in practice, will naturally make a physician suspicious of those which were introduced by false reasoning or superstition. Yet it must be owned, that some valuable remedies have sometimes been discovered in consequence of absurd theories. Another advantage attending a knowledge of the history of physic, is its bringing us acquainted with some [89] efficacious remedies which time and other accidents had thrown into disuse.

The change of manners, and the variations of our speculative systems of physic, have, in some degree, contributed to the less general use of certain bold but successful remedies employed by the ancients; as might be exemplified in the case of cauteries, the application of various exercises, of frictions and of unctions, and in other instances. The history of medicine likewise shews us, how the revolutions of time bring back really the same fanciful hypotheses, which, only by a change of terms, have been repeatedly obtruded on the world.

Although the progress of medicine, since the age of Hippocrates, has indeed been slow, considering the number and abilities of its professors, yet it has made considerable advancement since that time. The history of physic shews how it has been gradually improved by accidental [90] discoveries, by the rash attempts of empirics, by the accurate and faithful observations of sagacious physicians, and by the sober and diffident reasonings of men of true medical and philosophical genius. Nor should it be thought, that even the most fanciful hypotheses that have prevailed in physic have been entirely useless. The zeal of supporting a theory however false, has given rise to some important experiments. Enthusiastic chemists, who boasted of a command over nature, and trusted to the efficacy of their own medicines, have sometimes performed surprising cures, and by such remedies as no physician would have ventured on. On the other hand, Stahl and his followers, who trusted almost every thing to Nature, have advanced the art by their diligent attention to the history of diseases, and to the operations of Nature in performing the cure. [91]

I am at a loss what advice to give you in relation to the study of mathematics, because I distrust my own judgment on this subject. I am afraid I am partial to a science to which I had a kind of innate and heredetary attachment, and which was the business and pleasure of my early days. An acquaintance, at least, with the elements of this science is certainly necessary, if we would make any progress in natural philosophy; and I have already shewn how intimately that science is connected with a thorough knowledge of the animal œconomy. The application of mathematics to medicine, towards the end of the last century and the beginning of the present, was productive of some good consequences. Among others, it contributed to banish the false hypotheses of the Galenical and chemical sects, and that scholastic jargon which had involved physic in unmeaning verbal altercations. It introduced a more liberal spirit of enquiry into every branch of [92] medicine; a greater attention to experiments and observations, and a greater degree of clearness and precision in medical reasoning. I acknowledge, however, that this study has often been abused. Many of the mathematical physicians were unfortunately persuaded that all the phænomena of the animal oeconomy were explicable on mechanical principles alone: But the impossibility of applying these upon some occasions, and the too hasty application of them upon others, often led to very false conclusions. Indeed, any person accustomed to the accuracy, perspicuity, and elegance of geometrical reasoning, must see with some indignation the parade of mathematical language, and the prostitution of the word *demonstration*, so frequently misapplied. From what I have said, you may perceive that I do not recommend the study of mathematics as leading directly to any important discoveries in medicine, but from a persuasion that, besides its subserviency to natural philo-[93]sophy, it has a tendency to quicken the invention, to open the mind, and to accustom it to a habit of close and accurate reasoning. But, let me caution you against entering too deeply into this most bewitching of all studies, which will probably divert your attention from the main ends of your profession. Let me also desire you to guard against its leading you to a disposition to scepticism, and suspense of judgment in subjects that do not admit of mathematical evidence. Remember that habits of nice discrimination, though frequently useful, are sometimes incompatible with the business of common life, and of your own profession.

I hope I have no need to recommend to you a thorough acquaintance with the Latin tongue. A physician's reading must be confined within very narrow bounds, who is unacquainted with what has been the universal language of the learned in Europe for so many ages, and [94] which serves to communicate their sentiments, from one nation to another, so easily and so quickly. The interests of learning will very soon suffer by its disuse, and by the present mode of authors writing in their own native language. But I must here take notice of an error, which they who value themselves on their knowledge of Latin are apt to fall into, and which has contributed, more than any thing, to this growing evil. What I mean is, too great anxiety about classical purity, and elegance of expression. The intention of language is to convey our ideas with clearness, force, and precision. If these can be joined to a style truly classical, it is a great additional beauty; but, from the numerous improvements made by the moderns in all the arts and sciences, there have arisen many ideas and objects, which the Roman classics could have no expressions for, because they did not know them. An author, therefore, who has occasion to express these ideas, is under a neces-[95]sity of latinizing words of his native language, in order to express his meaning, or of adopting Latin words used only by authors of inferior note. If he is determined to use no phrase but what is strictly classical, he must, on many occasions, either suppress what he wished to say, or he must lose that honour he gains by elegance of diction in the more substantial points of energy, precision, and perspicuity. We have the peculiar advantage in our profession of having a standard in Celsus for purity and elegance of medical Latin; but there are still a variety of medical ideas, arising from the improvements in the science, to be expressed, which neither Celsus, nor any Roman writer, could be possessed of.

I must here recommend to you the study of that copious, expressive, and harmonious language, the Greek, in which some of our oldest, and some of our best authors have written; parti-[96]cularly Hippocrates, the father and founder of medicine. Almost all the medical terms of art are Grecian; a knowledge, therefore, of that language must evidently facilitate your progress in the profession. Besides, it is unbecoming a physician, who pretends to a liberal education, to be in the daily use of terms to whose original he is a stranger.

The necessity of a knowledge of the French is apparent. Almost all the authors of that nation, many of whom are very valuable, write in their own language; it is likewise become so universal in Europe, that every gentleman who travels must necessarily make himself master of it.

It may appear at first view superfluous to recommend an attention to your own language. But it is well known, that many physicians of real merit, have exposed themselves to ridicule by their ig-[97]norance of, or inattention to, composition. It might be expected, that every one who has had the education of a gentleman, should write his native language, with at least grammatical exactness; but even in this respect, many of our writers are shamefully deficient. Elegance is difficult to attain; and, without great taste, dangerous to attempt. What we principally require in medical writings, is perspicuity, precision, simplicity and method. A flowery and highly-ornamented style in these subjects is entirely out of its place; and creates a just suspicion, that an author is rather writing from his imagination, than copying from Nature. We have many bulky volumes in medicine, which would be reduced to a very small compass, were they stripped of their useless prefaces, apologies, quotations, and other superfluities, and confined to the few facts they contain, and to close inductive reasoning. – What I would principally recommend to you in every species of medical writing, next to [98] a simple and candid history of facts, is a strict attention to method. I am no admirer of that display of system and arrangement, so remarkable in some writers, who split their subject into endless divisions. This may strike a young reader, not accustomed to such kind of writing, with an high opinion of the author's ingenuity and accuracy; but in general it is a mere deceit. It is a mode of writing easily attained, and was in the highest perfection when the scholastic logic, which consisted rather of nominal than real distinctions, was held in admiration.

I must however observe, that the composition of a book of science intended only for the perusal of our own faculty, and that of one addressed to the public in general, upon a subject in which they are particularly interested, and which they are capable of understanding, may, and ought to be different. In the first, the qualities I just now mentioned are alone [99] necessary. The intention here is only to investigate truth, to communicate discoveries, to relate new facts, or to exhibit those already known, in new lights and combinations. The book is addressed to men who must be supposed willing and ready to give it a fair and impartial review. But in a medical work addressed to the people, the design may be to make them not only think rightly but act properly, not so much to instruct as to reform. It must therefore be written in such an agreeable manner, as may entice them to read it. Not only must the subject be stript, as much as possible, of all technical terms, and made perfectly level to their capacities, but the author may be allowed to paint the truths it contains in lively colours, and in such a manner as to warm the imagination, and interest the feelings. Without this, a few speculative men may afford it a cold approbation, but the book will not be generally or well received.

Its merits, in regard to the facts it contains, and the [100] justness of the reasoning from those facts, can perhaps be properly decided only by the faculty: Its merits as a composition must be estimated by the public approbation, and the good effects it produces.

It would require too much time to enumerate all the qualifications that might be deemed ornamental to a physician. In general, there is no reason why he should be excluded from any amusement, or any accomplishment, that becomes a gentleman. On the contrary, these give an agreeable relaxation from the severer studies and fatigues of his profession; they render his conversation more cheerful and entertaining; and, instead of that aukward pedantry, by which modern men of learning have sometimes been distinguished, they diffuse an ingenuous and liberal air over his whole behaviour. [101; 102 is blank.]

LECTURE IV

General views and principles to be attended to in the investigation of Nature. – Advantages attending the study of Nature – favourable to religion. – Natural history of man, what it includes. – Every event in nature happens in consequence of general laws. – These laws how ascertained. – Original principle of belief in mankind. – Experience how attained. – Evidence of our senses sometimes fallacious, sometimes deficient. – Consequence of trusting to a limited experience. – Reasoning by analogy – deducing general principles by induction from particular facts. – Errors we are led into by our impatience to ascertain these principles. – Deception from ima-[103]ginary analogies. – Advantages of philosophical diffidence. – Necessity of reasoning and of establishing general principles, particularly in medicine. – State of the controversy between the Empirics and Dogmatists.

The works of Nature are of infinite extent and variety: but, amidst all this variety, there is, as I have already remarked, such an intimate connection, that no one part can be thoroughly understood by studying it entirely detached from the rest. In our enquiries into the various branches of the works of Nature, there are certain general views, and certain general principles of investigation, to be particularly attended to. The general views to be attended to, in the study of Nature, respect, 1. The advantages it brings to individuals. 2. Public utility.

1. The advantages to individuals that attend enquiries into Nature, are suffi-[104]ciently obvious. These inquiries give exercise to many of the active powers of the mind; they gratify curiosity, the love of truth, and of whatever is great, beautiful, or wonderful: principles deeply implanted in human nature.

2. In regard to public utility, they promote all the useful and elegant arts, all the arts that tend to the happiness and ornament of human life. A profound knowledge of Nature extinguishes pride and self-conceit, by rendering men more deeply sensible of their ignorance, their errors, and the very limited state of their faculties. It is favourable to the interest of religion, by exhibiting the most striking proofs of the infinite wisdom, power, and benignity of the Supreme Being, who supports this wonderful frame of things, by laws, often, indeed, unsearchable in their nature by human wisdom, but steady and uniform in their operation, and admirably fitted to promote the happiness of his creatures. Such a know-[105]ledge must impress every heart endued with the least portion of sensibility, or not strangely perverted, with that awful veneration, that love and gratitude to the Divinity, that submission to his providence, and that reliance on his goodness, which alone constitute true devotion. It has been imagined by some, that very extensive knowledge leads to Atheism; but there is not the least ground for such a suspicion. A little learning is, indeed, a dangerous thing to a weak and conceited man, who, from a superficial acquaintance with second causes, is apt to overlook the First and Great Cause. But to a sound understanding, extensive knowledge is the truest teacher of humility; it shews how often men are deceived in their supposed acquaintance with second causes; and that, even where many of these are clearly ascertained, yet, in tracing the chain that connects them, the most acute and profound genius must stop somewhere, and at last refer them to a supreme intelligent First [106] Mover. While we attempt, however, to clear philosophy from the charge of impiety, a very important distinction must be attended to. I will venture to maintain, that those philosophers have been the firmest supporters of religion, who have employed their genius and application in the investigation of the works of Nature, and whose views in science have been grand and extensive. Among a multitude of examples I could bring to prove this assertion, I shall mention only three of our own countrymen, Lord Bacon, Mr. Boyle, and Sir Isaac Newton. Those philosophers, on the other hand, who have been the most distinguished propagators of Atheism, have been men not much acquainted with the works of Nature, who searched for truth in

their own little minds, not in the great world without them*; [* "Bacon"– at bottom p. 107] men who, in regard to science and the useful arts, have either neglected them altogether, instead of pro-[107]moting them by observation and experiments, or corrupted them by metaphysical subtleties, often indeed ingenious and plausible, but which have never led to any useful discoveries or improvements.

The system of principles merely hypothetical, fabricated by these men, and sometimes supported with much subtlety and some fancy, with a view to disprove the evidence of a Divine Providence, of the immateriality of the soul, and of a future state of existence, shews a wonderful perversion of the human mind. Surely, if we ever chuse to soar into the regions of Fancy, it should be from a desire to amuse, to mend the heart, to warm the imagination with pleasing prospects of the dignity of human nature, of Providence and of futurity; and not with a view to degrade our nature, to infuse suspicions upon the most interesting subjects, and to throw a damp upon every heart possessed of the common feelings of humanity. [108]

There is no branch of natural knowledge so useful or interesting, as that which relates to the human species; which is evident, when we consider that it includes,

1. Medicine, or the art of preserving health, of prolonging life, of curing diseases, and of making death easy.

2. The arts of improving the different faculties of the human body; as strength and agility, rendering us superior to pain, cold, hunger, and the many other evils mankind are subjected to.

3. The preservation and improvement of beauty.

4. The laws of union between the mind and body, and the mutual influence they have upon one another. This is one of the most important enquiries that ever engaged the attention of mankind, and almost equally necessary in the sci-[109]ences of morals and of medicine. This last comprehends,

(a) The doctrine of the preservation and improvement of the different senses, external and internal, the memory, imagination, affections, and judgment.

(b) The history of the power and influence of the imagination, not only upon the mind and body of the imaginant, but upon those of other people.

(c) The history of the several species of enthusiasm.

(d) The history of the various circumstances in parents, that have an influence on conception, and the constitution and characters of their children.

(e) The history of dreams, with a view to our acquiring a power over them.

(f) The history of the power and laws of custom and habit. [110]

(g) The history of the effects of music, and of such other things as operate upon the mind and body, in consequence of impressions made on the senses.

(h) The history of natural signs and language, comprehending the doctrine of physiognomy and outward gesture.

(i) History of the power and laws of the principle of imitation.* [* "Bacon" – at bottom p. 111]

I mention these heads only as a specimen, and not as a full enumeration of the many important articles contained under the natural history of the human species. I mention them as examples of the general views to be regarded in our investigation of Nature, and essentially connected with the science of Medicine; but have taken no notice of the enquiries that relate to man in his moral, political, or religious capacity, as being foreign to my profession. [111]

I proceed now to lay down certain general principles, which require our attention in the investigation of Nature, and shall endeavour to apply them more particularly to the science of Medicine. When we look around us, we find objects connected together in a certain invariable order, and succeeding one another in a regular train. It is by observation and experience alone, that we come to discover this established order and regular succession in the works of Nature. We have all the evidence which the nature of the things admits of, to persuade us that nothing happens by chance: nay, we have every reason to believe, that all events happen in consequence of an established and invariable law; and that, in cases similar, the same events will uniformly take place.

Here I must observe, that, antecedent to all reasoning and experience, there is an original principle implanted in the human mind, whereby it is led to a be-[112]lief that the course of Nature is regular. In consequence of this principle, whenever a child sees any event succeeding another, he has an instinctive persuasion, that the same event will succeed it afterwards in the same circumstances. This persuasion does not flow from any connection he sees between the cause and the effect, nor from experience, nor from reasoning of any kind. So ardently do we desire to find every thing that happens within our observation, thus connected with something else, as its cause or occasion, that we are apt to imagine connections upon the slightest grounds; and this weakness is most remarkable among the ignorant, who know least of the real connections established in Nature. – Credulity seems likewise to be an original instinctive principle, by which we are disposed to believe, prior to experience, not only the language of natural signs, but also the language of artificial signs, as soon as they come to be understood. Hence the credulity so remarkable in [113] children. They at first believe every thing that is told them to be true; and it is experience alone which teaches them to restrain this natural propensity. Dr. Reid has treated this subject with great acuteness, in his ingenious Enquiry into the Human Mind.

We obtain experience, either by the evidence of our own senses, or by the testimony of others.

1. The testimony of our senses, though generally considered as one of the highest degrees of evidence, is often fallacious, and often defective. The sensations excited in us, in consequence of impressions made on our organs of sense, depend on the following circumstances:

On the state of the medium through which the communication between the objects and the organs of sensation is made, e.g. the state of the air, when we judge of visible objects. [114]

On the state of the organs of sensation themselves, every one of which may be vitiated in a variety of ways.

Our unassisted senses often fail us on account of the subtlety or minuteness of bodies, too quick or too slow motion, the object being too common; and many other causes.

Although the impression is properly made on organs that are in their sound state, yet the ideas conveyed thence to the mind may be so varied and modified by the imagination, as entirely to mislead the judgment. Thus every part of natural history, and medicine above all others, is crowded with facts, attested by eye-witnesses of supposed veracity, which, notwithstanding, had never any existence but in their own imaginations.

From a failure likewise of memory, and from imagination assuming its place, we believe, upon the supposed evidence of [115] our senses, although in fact we never had such testimony [substitute 'evidence' in errata]. We likewise often mistake an opinion, or an inference of the understanding, for a fact established by the evidence of sense; for example, when we judge of the real magnitude of objects by their appearance. But although, in some particular circumstances, our senses may be fallacious or deficient, yet we are led, by an irresistible instinct in our nature, to trust to them. All experimental knowledge proceeds upon this principle; nor can we make one step in life without it. The very methods we use to discover the fallacies and deficiencies of our senses, presuppose such a necessity to yield to their evidence, because the appeal must always be made to this evidence.

2. The experience which we trust to from the testimony of others, is liable to the same imperfections with our own, and often to the additional inconvenience of our uncertainty with regard [116] to the accuracy or veracity of our authors.

Having examined the sources of experience, I shall now proceed to consider the manner in which mankind agree in applying it. I have already remarked, that men naturally believe, that what they have seen to happen in one case, will happen again in the same circumstances; and that the same causes will always produce the same effects. This is equally true, both with regard to the philosopher and the peasant. The only difference between these two consists in this; the peasant concludes two cases to be precisely alike, because they resemble one another in their obvious appearances; the philosopher, on the other hand, from a more enlarged experience, and more accurate observation, does not so easily trust to appearances; he is aware of the various sources of deception, and therefore examines every minute and latent circumstance, before [117] he ventures to form a judgment; and the difficulty of ascertaining, with precision, the exact similarity of cases, makes the true philosopher extremely sceptical in drawing conclusions of what will happen, from what has happened. An African, who has seen water in a variety of circumstances, but still retaining its fluidity, concludes, that fluidity is essential to water, and will not believe travellers when he is told, that, in certain parts of the world, water often appears in a solid form. His mistake does not proceed from trusting to experience, but from thinking he had sufficient experience, when in reality he had it not. All that he could justly infer from his experience was, that water, in all the circumstances under which he had seen it, would remain fluid. But water exposed to a degree of cold sufficient to congeal it, was a circumstance in which he never saw it; therefore his experience could never tell him, what effect that cold would have [118] upon the fluid, whenever it came to be exposed to it.

Although facts afford the only solid foundation for genuine science, yet, when we consider them as unconnected with other facts, they convey but little instruction. The phænomena of Nature are infinite; but the capacities of the human mind, and particularly the memory, are limited. If these phænomena, therefore, were not reducible to certain general principles or laws, our experience of particular facts could do us but little service. But there is a strong propensity in the mind, to be delighted with analogies; to compare and connect facts that resemble one another, and, by this comparison, to reduce them to certain general principles, to apply such general principles to account for other effects, or to direct us in the production of them. The business of true philosophy is, to avail itself of this natural propensity, to discover these connections, and reduce [119] them under certain general rules or principles, called *laws of Nature*. Our inclination to reduce particular facts to general laws appears from the anxiety which men shew to discover the cause of any uncommon event. The discovery of this cause infers no more, than the finding out that law of Nature, by which the event is produced. In our enquiries into Nature, after we have arrived at the knowledge of some general laws, by an accurate comparison and arrangement of observations, we may, by comparing these laws together, discover laws still more general; and thus, by a slow and cautious induction, we may advance to a knowledge of the most general laws that regulate the system of Nature. But many obstacles concur to prevent the establishment of philosophy upon this solid foundation; some of which I shall endeavour to explain.

1. The * [*"Bacon de Augmentis Scientiarum"– at bottom p. 120] impatience of men to reduce all knowledge, and to refer all events [120] to certain general laws, makes them unwilling to submit to a slow, but sure, method of investigation. They attempt a shorter way of discovering those laws, in which they are misled, either by a false reasoning from imaginary analogies, or by supposing the laws of Nature to be fewer and simpler than they really are. The consequences of which are, the hasty reduction of the sciences into imperfect and erroneous systems.

2. The pleasure that men have in discovering analogies, makes them often imagine resemblances between things, where in truth there are none, or none of any consequence. Arguments from analogy very readily present themselves to a warm imagination, while more direct and conclusive arguments, drawn from observation and experiments, often require painful attention and application; though, after all, they may prove insufficient to establish the wished-for principle. I shall readily acknowledge the [121] usefulness of analogies; they often facilitate the conception of things, which, without their assistance, could not easily be comprehended. It is likewise by reasoning from analogies, that we are sometimes led to anticipate many useful principles and discoveries. But we ought never to

acquiesce in analogies, when we can obtain more direct evidence; as all that analogies can lead to, are but probable conjectures, commonly called Theories.

3. There is a certain intoxication, usually attending the supposed discovery of general principles in science, or of useful inventions in the arts, which renders men of warm imaginations blind to every difficulty that lies in their way, and often induces them even to suppress such difficulties. The concealment of facts that contradict a favourite hypotheses, is not always owing to want of candour. Sometimes the author does not mention them; because he does not see them; [122] sometimes he disregards them; and sometimes he conceals them, from the fear of creating a prejudice against what he thinks an important discovery. Every true philosopher, however, will be particularly diffident of himself in this respect; and whenever he gets a glimpse of a theory, will immediately set his invention at work, to contrive every experiment that can produce a direct evidence, either of its truth or falsehood.

This philosophical diffidence is so far from discouraging, that it greatly promotes the investigation of causes and general laws. A state of suspense is always disagreeable; and the uneasiness it gives, becomes a powerful incitement to such further enquiries as may remove it. A zealous attachment to theories, may not only lead into dangerous mistakes, but, by betraying men into a false security, cut off every motive to further enquiry. It is not a true philosophical scepticism, nor a low opinion of our [123] present knowledge, which checks the spirit of enquiry into the laws of Nature; it is a mean opinion of the human powers, which effectually chills the ardor of genius, and blasts all grand and extensive views of improvement.

In works addressed to the heart, that coldness and severe precision so necessary in the investigation of truth, have no place: imagination there is in her proper element, and the loosest and wildest analogies may be often admitted with propriety. A philosopher may read a fairy tale with great delight, without giving the least ground for calling in question his taste or understanding; but it reflects severely upon both, if he reads with the same pleasure a philosophical investigation, not founded in observations and experiments, but in the vagaries of a lively imagination; unless he be sensible of its being a romance, and only allow himself to be charmed [124] with the spirit or elegance of the composition.

4. There is, in matters of this kind, a species of self-deceit, which deserves particular notice. We often find those people inveighing bitterly

against theories and hypotheses in philosophy, who are themselves (perhaps without knowing it) notoriously addicted to them. This is remarkably the case with medical writers, who commonly decry all reasoning and principles in physick which differ from their own, as idle theory; and frequently declaim against theory, in such a way as might seem to condemn all reasoning and investigation of causes and principles as useless and delusive. But it should be considered, that we cannot advance a step in the pursuit of knowledge without reasoning. In every useful experiment, and especially in conducting a train of experiments, we must employ our reason; there must be some point in view, some anticipa-[125]tion of a principle to be established or rejected, and reason must determine all the circumstances to be attended to in making every observation, or experiment, with a view to discover the truth. Without reasoning, or without trusting to certain principles, either established or rendered probable, we could never be benefited by experience, because we could never transfer it from the case we have seen, to the case immediately before us. For instance, I have a patient in an intermitting fever, which I propose to cure by the Bark. I shall suppose I have cured five hundred patients by this medicine formerly; but yet I know I never cured one whose condition, in respect of age, temperament, and other particulars, exactly corresponded. If, therefore, I give this medicine, I must reason upon this principle, that the Bark will universally cure this disease, notwithstanding some difference of circumstances. But this is a principle of which I have no direct and conclusive [126] knowledge, but one which I have adopted by a probable reasoning from analogy: and in reality, it is not universally true, though physicians must proceed upon it in their practice, till future observation shall ascertain the exceptions. Boerhaave, Hoffman, Stahl, and other systematic writers, exclaim against theories, meaning one another's theories; for each of them explain, though in a different, and often opposite manner, the proximate cause of every disease they treat of, and the mode of operation of every remedy they prescribe, upon principles too hypothetical. Even Sydenham, though reputed a purely practical writer, is full of hypothetical reasoning, which, however, had not the usual effect of making him less attentive to observation; and, indeed, his hypotheses seem to have sat so loosely about him, that either they did not influence his practice at all, or he could easily abandon them whenever they would not bend to his experience. [127]

It should seem, upon the whole, that all physicians must reason; and that the only difference among them consists in this, that some reason better than others. Some, for example, search into the causes of diseases, and the effects of remedies. Deeply sensible of the difficulty of the enquiry, and the various ways in which they may be deceived, they collect and arrange all the facts relating to the subject, when they have got a remote view of a leading principle, they attempt, by experiment, to bring a direct and conclusive proof of its existence. If the proof turn out against it, they see, and candidly acknowledge their mistake; if the case does not admit of a direct proof, they consider [substitute 'think' for 'consider' in errata] their principle as only more or less probable, but never relinquish the pursuit. These, I think, have a just claim to the title of *rational physicians*. Others, upon the foundation of a few facts and vague analogies, form hypothetical principles. A creative imagination supplies materials, where [128] they seem wanting; they employ their ingenuity to strain facts into a correspondence with them, and such as will not bend to their purpose they reject, under pretence of their being false or incredible. In practice, they neglect particular observations; because they think their general principles so well established, as to want no confirmation. Such men assume the title of Rational Physicians. But surely every system-builder, who has classed himself among the rationalists, cannot have a claim to this title; because many of their systems are different and contradictory.

But from the days of Serapion, founder of the empirics, to the present time, there has been constantly a division of physicians into two sects, one pretending the strictest regard to observation and experience, but stigmatized by their opponents as quacks; the other assuming the name of rational or dogmatical phy-[129]sicians, but accused of being contemners of experience, and of being attached to imaginary hypotheses, either wholly inapplicable to practice, or corrupting it with errors. This division appears to me to have essentially hurt the interests of medicine; and as many students are apt to inlist themselves on the one side or the other, who can have no just idea of the nature of the dispute, I shall endeavour to lay it open to its source, and shew how much men have been deceived and perplexed on this subject by the ambiguity of language. But it may be proper to premise a general view of the state of medicine before this controversy existed.

There are no traces of any regular system of medicine before the days of Hippocrates. The practice before his time seems to have been merely empirical; that is, founded on real or imaginary experience of the effects of remedies in particular diseases, but without any pro-[130]per regard to their symptoms or causes. It was, besides, as appears by the earliest accounts we have of it in Egypt, confined to the priests, which was generally the case among the most ancient nations, who concealed it as a mystery, interwove it with their religious superstitions, and exercised it with much artifice. Other inconveniences arose from the science being, for many ages, confined to a particular family in Greece, the descendants of Æsculapius.

It is evident that philosophers who were not of this family, and who began to study medicine as a useful branch of natural philosophy, were the first who introduced into it a more enlarged spirit of observation and reasoning. The most distinguished of these was Pythagoras; who, with a penetrating genius, enquired, with unbounded curiosity, into every part of Nature. His warm imagination led him to a belief in genii, in magic, in visionary harmony and powers [131] of numbers, which tinctured all his philosophy, and consequently his physic, with which it was connected. Succeeding philosophers further corrupted medicine, by distorting every branch of it with the most chimerical hypotheses.

In this condition it was found by Hippocrates. His sagacity discovered to him the necessity of correcting these abuses; and he set about new-modelling the art, and placing it upon the sure basis of observation. Hence he is said to be the first who separated the study of medicine from that of philosophy. But, as he likewise endeavoured to establish general principles from particular observations, he is considered as the father of the rational or dogmatical system. It appears, however, from his works, that some part of the prevailing philosophy, and even of the superstition of the times, still adhered to him: but, on the whole, his reasoning is more just than could be expected, considering the then low state of ana-[132]tomy, and other sciences connected with medicine.

For some centuries after Hippocrates, medicine seems to have made no progress. Two of the greatest men of antiquity, Plato and Aristotle, concurred, though in different ways, to check its improvement, not only during that period, but almost down to our times.

Plato, whose writings are distinguished by the purest Attic elegance, will always be considered as one of the fathers of polite literature. But to view him as a natural philosopher, he must be accounted a corrupter of almost every branch of it, and particularly of medicine. In his Timaeus, he has given a specimen of his sentiments relating to the animal œconomy, which are the mere effusions of a heated imagination. The sublimity of his genius, which attempted to grasp the whole creation, and his irresistible eloquence, captivated all succeeding philo-[133]sophers whose imagination was superior to their judgment, by withdrawing their attention from the study of Nature, under the specious name of Contemplation. So that many of his wildest theological ideas have been introduced into our systems of physic, as well as into those divinity.

Aristotle possessed a most acute and comprehensive genius. His writings, in many branches of knowledge, are deservedly held in the highest esteem, both for ingenuity and soundness of reasoning. But I am only to speak of such of his works as relate to natural philosophy and medicine. The writings of this philosopher, though he does not treat professedly of medicine, have had a more extensive influence over it than those of any physician whatever. His philosophical principles were similar to those of Plato, were hypothetical and visionary, but supported with more plausible arguments, and founded on a more extensive [134] knowledge of Nature. As his principles were adopted by Galen, almost all the systems of physic, till those of the seventeenth century, were more or less derived from them. But the prejudice he has done to physic has not so much arisen from his introducing into it false principles, because time must have discovered this abuse; it has arisen from his having corrupted the true spirit of philosophical investigation. Under pretence of teaching men to reason with clearness and precision, Aristotle, or perhaps rather his followers, the school-men, stopped the progress of useful knowledge, by diverting the attention from experience and observation, and engaging it in the pursuit of useless subtleties, which professed to penetrate into the deepest recesses of Nature, but in reality ended in nothing but in useless jargon.

Different modifications of the doctrines of Plato and Aristotle, and of some others which time has buried in [135] oblivion, confounded medicine with many absurdities, and involved it in disputes with which it had no concern. In this situation, it was found by Serapion, who maintained that this philosophy was foreign to the art of medicine, the practice of which he confined to experience alone. He deemed it unnecessary to inquire into any causes of diseases, but such as were evident, and therefore rejected anatomy, the dissection of morbid bodies, and all enquiries into the remote and latent causes of them. However absurd such a conduct would justly appear to us in the present improved state of medical knowledge, in those days it was plausible. Physicians were then ignorant of every branch of natural philosophy connected with their profession, as well as of anatomy, physiology, and chymistry. If they had been sensible of their ignorance in these matters, their reasoning might have been defective, but it would not have been erroneous, because it would have stopped whenever they wanted facts upon [136] which they could proceed. But this has seldom been the practice of philosophers or physicians in any age: wherever observations have failed, fancy has always supplied their place; so that materials have never been wanting to establish for a time, any hypothesis whatever. I shall mention as a specimen of the medical philosophy of ancient times the following doctrine from Plato.

The first form which the elementary particles of matter received, was believed to be triangular. From the different sizes and positions of these triangles, were produced the four elements, fire, air, water, and earth. Fevers were thus accounted for. If the fire exceeded, continued and ardent fevers were produced; if air, quotidian intermittents; if water, tertians; if earth, quartans. A method of cure in these diseases was laid down, supposed to be correspondent to the proximate causes just mentioned, and in many respects fully as absurd. It is a lesson [137] of humility to find, that the human understanding, in a most enlightened nation, among men of the most distinguished genius, could be weak enough to embrace such chimeras for truths.

Serapion had several followers among the ancients, eminent for their abilities, who were distinguished by the name of *Empirics*; But there have been so few among the moderns who had any pretensions to learning or genius, that have openly professed themselves of this sect, that the name of Empiric is now used as a term of reproach, and only applied to illiterate quacks. But, tho' all physicians regularly educated, disclaim the name of Empiric, yet, in effect, the ancient distinction between empirics and dogmatists is continued, in other terms, even to the present times. But it appears to me, that both parties have been to blame, not only in regard to their conduct as physicians, but in the loose manner in which they have carried on the dispute. [138] In order to prove this, I shall endeavour to settle the meaning of some words that often occur in this controversy, the ambiguity of which tends to perpetuate it.

An *empiric* properly signifies a physician who regards experience, and who is directed by it in his practice. In this sense it is creditable to be an empiric. – Experience is surely the foundation of all knowledge in physic; the ultimate appeal must be made to it, and whatever assertion contradicts experience or facts, ought to be rejected as false. The appellation, *Empiric*, however, is generally applied to one who, from observing the effects of a remedy in one case of a disease, applies it to all the various cases of that distemper. But the names given to diseases are in reality names annexed to a certain number of symptoms, rarely exceeding three or four. The same name, therefore, is given to all the cases where these symptoms occur; or, in other words, these cases are referred [139] to the same genus of diseases, although, in many other respects, they are very different, and require a different treatment. Thus, under the genus of small-pox, are comprehended several species, varying more from one another than an inflammation of the lungs from a rheumatism. There can therefore be no universal medicine for every case of the small-pox, or indeed for any other disease; because, though they may agree in the few symptoms which give the name, yet they may differ in regard to others of more consequence. The application of a remedy in a disease must likewise be regulated by the various remote and proximate causes producing it, by the constitution of the patient, by his age and habit, by the season of the year, the climate, and other circumstances.

These truths are so well known to every person in the least acquainted with physic, and one should imagine are so ob-[140]vious to common sense, that it is surprising how any man could be so ignorant or impudent as to recommend an universal specific for any one disease, and equally surprising how any man of the lowest understanding, should give the least credit to such an assertion.

From what I have said, it appears, that empirics, notwithstanding their pretensions of relying upon experience alone, have in truth abandoned it.

There is an experience, indeed, which, however extensive, does not render a physician more sagacious, or more successful in his practice, because it is not attended with the necessary observation. Some set out with a belief in the infallibility of certain principles, and of remedies resulting from those principles, in the cure of diseases. These they are sure to apply, as soon as they have fixed the name of a disease. They give no attention to the distinction to be made between the cases, [141] where their remedies have, or have not succeeded; they never vary their practice, nor listen to any proposed improvement, and consequently cannot profit by any new discoveries.

The fate of medicine and that of agriculture have in this respect been similar. Within these last thirty years, more real knowledge has been acquired in these two sciences, and more facts ascertained, than in many preceeding centuries, while at the same time the uncertainty, and even falsehood, of many of their principles have been demonstrated. Yet what has been done, serves principally to shew physicians and farmers how much they have been hitherto deceived in their supposed experience, and to give them some idea of the extent and difficulty of their several professions.

From the manner in which empirics in all ages have conducted themselves, it is not surprising that their writings [142] have tended so little to the advancement of the art; and that, on the contrary, they have had the greatest share in encumbering it with the many falsehoods under which it has laboured so long, particularly that important branch which relates to the effects of medicines. – It has been pretended, that such empirical books as I have alluded to, may be useful to those who are not bred to the profession, and who wish only to acquire some knowledge of the practical part of physic. But this is so far from being the case, that these are the only people to whom such books can be dangerous. A physician of real knowledge and practice may draw instruction, or catch hints from facts related in an imperfect manner, which will either be useless, or tend to mislead others who have not these advantages. To such, all the circumstances relating to the exhibition of a remedy can never be too distinctly specified. I shall give an example in the case of one, which I shall suppose recom-[143]mended as almost infallible in the cure of a head-ach. How many questions occur here? – In what kind of head-ach has it been found serviceable? Did the pain arise from any affection within the head itself; from a congestion of blood; from a collection of water; from inflammation of the brain, or of its membranes, or did it proceed from a disorder in the alimentary canal; from acidity; from any putrid matter, or collection of viscid phlegm in the stomach? Was the head-ach attended with fever; and with what kind of fever? Was it the consequence of sudden exposure to cold or heat, or was it the effect of ebreity, of wounds, or of other external violence; or of any strong affection of mind? A variety of other questions might be properly proposed in this example; and unless these were distinctly answered, the practice is defective. In many cases of a head-ach, the remedy could not be applied with any probability of success, and in others its application might be [144] attended with danger. Happily for mankind, by far the greatest part of those medicines which have been celebrated under the name of Specifics in particular diseases, are either so
very insignificant, or at least so trifling in the doses in which they are administered, that they may be given with safety in any case whatever.

Having attempted to shew how little medicine has been benefited by the labours of the Empirics, I shall advanced by their opponents, who assume the name of Regular, Rational, or Dogmatical Physicians.

The term, *dogmatic*, in its original sense, implied only one who endeavoured to reduce his knowledge of diseases to certain principles. It came afterwards to be adopted by physicians, who, from weakness and vanity, pretended to practice, from a knowledge of the pro-[145]ximate causes of all diseases, and of the mode of operation of all remedies. But now, in common language, the term, Dogmatical is generally used in an unfavourable sense, being applied to one who is conceited, dictatorial, and obstinately attached to particular opinions.

The complaints against those who assume the character of rational and dogmatical physicians, have been too loud and too frequent to be entirely without foundation. They have been accused of neglecting observation; of withdrawing the attention from the useful and practical part of medicine; of engaging the mind in disquisitions foreign to the main ends of their profession; of corrupting every branch of medicine, by false reasoning and ill-founded hypotheses; and of disguising, suppressing, and even forging facts, in support of their visionary systems. I am sorry to say, that the history of medicine in all ages sufficiently justifies these charges; but at the same time [146] it must be observed, that they have been carried to an extreme. Some people, not satisfied with railing at all hypotheses and theories, exclaim against all reasoning in physic as manifestly tending to mislead us. But here we must endeavour to ascertain the meaning of some terms made use of in these complaints.

Reasoning properly signifies the exercise of that power of the mind by which it infers one thing from another, or deduces conclusions from premises. Without the exercise of this power, we could neither act in the common affairs of life, unless when impelled in particular cases by instinct, imagination, or passion; nor advance a step in the investigation of truth, beyond self-evident principles. As then we must reason from the necessity of our nature, our business is only to take care that we reason justly. But false reasoning is not more common in physic than in law, in divinity, or in [147] the common conduct of life; yet no one ever insinuated, that we ought to abandon the use of our reason in any of these subjects.

The chief objection to theory in physic proceeds from an ambiguity of words. The theory of a science properly signifies the doctrine of the general established principles of that science, and is distinguished from the practical art of applying those principles to the uses of life. Thus, for example, the theory of navigation does not consist of [add 'such' in errata] hypothetical principles, but on such as are established on solid and unquestionable foundations, and is distinct from the practice of navigation, which consists in the application of those principles with propriety and facility; an art to be acquired by habit and experience alone. There ought to be the same distinction between the theory and practice of physic; but, by a perversion of language, the theory of medicine is represented by some as a pretended science, [148] consisting, in reality, of mere conjectures, and chimerical speculations. In consequence of this unfair representation, which I formerly endeavoured to refute, an unhappy opposition has been established between the theory and practice of physic, as if they were not only not essentially connected, but were ['were' deleted in errata] even at variance; as if the one was entirely a creation of the imagination, the other the result of sagacious observation and experience: Whereas, in reality, theory, in the proper sense of the word, is produced by practice, is founded on facts alone; and constantly appeals to them for its truth. The prejudice which many people entertain against hypotheses in physic, is likewise founded on the equivocal signification of the word. It is commonly confounded with theory, but hypothesis properly means the supposition of a principle of whose existence there is no proof from experience, but which may be rendered more or less probable by facts, which are neither numerous enough, nor adequate to infer its ex-[149]istence. When such hypotheses are proposed in the modest and diffident manner that becomes mere suppositions or conjectures, they are not only harmless, but even necessary, in establishing a just theory in medicine. They are the first rudiments, or anticipations of principles. Without these there could not be useful observation, nor experiment, nor arrangement, because there would be no motive nor principle in the mind to form them. Hypotheses then only become dangerous and censurable when they are imposed upon us for just principles; because, in that case, they put a stop to further enquiry, by leading the mind to acquiesce in principles which may as probably be ill as well founded. In this way they have done great mischief in our science; but one of the chief advantages of a regular education, and of studying medicine on a systematic plan, arises from its rendering students more capable of distinguishing between real facts, and inferences of the mind mistaken for them; between a visionary hypothesis, and a just theory. [150]

LECTURE V

Error in supposing the laws of Nature to be fewer and simpler than they really are. – Natural dispositions of men influence their literary character. – Exemplified in those of lively and warm imaginations, and in those who are calm, sedate, and discriminating. – Bad consequences of a fondness for the Marvellous. – Abuses in the study of natural history. – Causes that have retarded the advancement of the sciences. – 1. Inattention to their end, viz, the convenience and happiness of life. – 2. Useless subtlety which may be displayed in different ways – too scrupulous regard to arrangement. – Observations on the subject of arrangement. – 3. Credulity. – 4. Attachment to great names. – 5. Blind admiration of antiquity. – 6. Fondness of novelty. – 7. Hasty reduction of the sciences into systems. – 8. Too [151] great attention to elegance of language, or an affected obscurity of style. – The study of Lord Bacon's writings recommended.

I OBSERVED before, that in our enquiries into human nature, an impatience to acquire a knowledge of her laws, and a natural love of simplicity, lead us to think them fewer and simpler than they really are. The more we know, the more we discover the uniformity and simplicity of the laws of Nature, when compared with the vast extent and variety of her works; but still we must not imagine that they are confined within the narrow limits of our knowledge, or even perhaps of our comprehension. When by an extraordinary effort of genius, Sir Isaac Newton discovered that all the planets gravitate towards the sun, by the same law by which bodies on the earth gravitate towards its center, many phænomena, whose causes we were till then ignorant of, were explained by this sim-[152]ple principle. But it soon came to be misapplied to the explication of other phenomena, which were afterwards found to depend upon very different laws.* [*"Dr. Reid's Enquiry into the Human Mind." at bottom p. 153] Des Cartes founded his system of the material world upon two principles, the existence of matter, and a certain quantity of motion originally impressed upon it. These, however, were found insufficient; and it has been made evident, that we must also admit the principle of gravitation just mentioned, cohesion, corpuscular attraction, magnetism, electricity, and other powers, by which the particles of matter attract, and repel, each other. Even Sir Isaac himself was led by analogy, and the love of simplicity, to conjecture, but with singular modesty and caution, that all the phænomena of the material world depended upon attracting and repelling powers in the particles of matter. But we have now reason to believe that he [153] was deceived: for, even in the unorganized kingdom, the powers by which salts, crystals, sparrs, and many other bodies, concrete into regular forms, can never be accounted for by attraction and repulsion in the particles of matter; and in the vegetable and animal kingdoms, there are evident indications of powers, of a different nature from those of unorganized bodies. We are conscious of an internal principle, which feels, which thinks, and which seems to be the origin of animal motions. We are, in a great measure, ignorant of its nature; but we know, that it has laws peculiar to itself, and that, in consequence of its union with the body, certain effects are produced, which the laws of matter are not sufficient to explain.

We may here observe how the different dispositions of men influence their literary character. We commonly find those of a lively and warm imagination, most disposed to attend to analogies, in [154] which fancy often deceives them. Upon these they are too apt to establish general principles, and to be so much attached to them, as not to see the objections to which they are exposed. If, however, by any accident, their opinion of their principles comes to be staggered, they two [substitute 'too' for 'two' in errata] quickly relinquish them, while, perhaps, they may be, in the main, well founded, though embarrassed with some difficulties, which a little more patience and perseverance might have conquered. To such the world is sometimes indebted for useful discoveries: They are themselves often ruined by projects, from their neglect of some small circumstance necessary to their successful execution, which a man of inferior parts, afterwards observing, robs them both of the honour and profit of their inventions. I shall farther observe, that this vivacity of genius is generally attended with an impatience, that renders men incapable of a proper attention to facts and experi-[155]ments, and prevents their bringing any work to a conclusion.

There is a species of genius the reverse of the former, calm, sedate, and attentive to the differences of things seemingly alike; that watches the operations of more lively and inventive spirits, and too often exposes their mistakes to ridicule. There are, in truth, so few men of original genius, who strike out new paths in arts and sciences, that they should meet with every encouragement, particularly when they propose their opinions with modesty. Men who go often out of the common road, must sometimes go astray; but as they now and then make important discoveries, their errors ought to meet with indulgence. These two characters are sometimes united in the same person in different degrees. One may possess that warm and lively imagination, so peculiarly fitted for invention, and, at the same time, a clear, accurate, and sound judgment, that candid-[156]ly considers every objection to his proposed plans; and, according to the weight of evidence, can either reject them altogether, or preserve his mind in a proper degree of suspence, till their real merit is ascertained. This happy union of genius and understanding, which we so rarely see, constitutes a philosopher of the first rank and dignity.

In collecting a natural history, subservient to the arts, and to become the foundation of a useful natural philosophy, it is necessary to make a selection of facts, among the infinite number with which Nature presents us. Our views should be confined to those which, being compared and arranged, may lead to general principles. The history, therefore, of any monstrous production, which has nothing similar to it, serves only to gratify curiosity. Yet this principle of curiosity, and love of the Marvellous, is so prevalent among mankind, that all the *lusus nature*, are what principally attract their regard. [157] If an animal comes into the world with two heads, we have presently a minute description of the monster published all over Europe, though it is not a matter of the least consequence to the advancement of science. This love of the Marvellous is conspicuous in some medical writers. We find them full of extraordinary cases, such as have nothing similar in them, such as never happened before, and, consequently, such as will probably never happen again, [add 'described' in errata] with a tiresome minuteness; while the symptoms that distinguish some common diseases from others of a different nature, which resemble them, are far from being yet ascertained. I do not mean here to object to the recording of every extraordinary event, as some of them may tend to throw light on the laws of Nature, in her ordinary course of proceeding. I only mean to censure this extravagant attachment to prodigies, when it leads us to neglect enquiries of more general utility. [158]

The present fashionable taste for natural history, regards it more as an object of curiosity, than as the basis of a sound philosophy, subservient to medicine, agriculture, and the other useful arts. Every natural production is not only tediously described, but accurately delineated. Surely it is rather too much to bestow a folio upon the natural history of a frog, in which that animal is painted in a great variety of attitudes. In this manner of extending natural history, it is evident that books may be multiplied beyond number, without bringing any accession of useful knowledge.

The advancement of the sciences has been much retarded by the following causes.

1. One of the chief causes* [* "Bacon de Augmentis Scientiarum." at bottom p. 159] has been an inattention to the principal end of their cultivation; that is, public utility, or [159] what contributes to the convenience and happiness of life. Instead of this, most men have no other object in the pursuit of knowledge, than to gratify curiosity, than to give a variety to their amusements, or to serve the purpose of vanity and ostentation. Perhaps no science has suffered so much as medicine by the neglect of its true end, which, as I before observed, is to preserve health, to prolong life, and to cure diseases. It has, indeed, made the slowest progress of any of the useful and practical arts; not so much from any deficiency of genius, as from a misapplication of it; nor yet from want of learning, for no profession can boast of more men eminent in every branch of useful and polite literature. Physicians have not only successfully cultivated every science connected with their own profession, such as anatomy, botany, chymistry, and the various branches of natural history, but have often distinguished themselves as poets, mathematicians, and philosophers. Yet [160] how few physicians can we name, who either by their genius or industry, have advanced the practical part of their own profession! how many, on the contrary, could we name, who have corrupted it, by the theories of their own imaginations, and even checked the slow improvement, which time naturally brings to every art founded on observation and experience! But the reason why medicine has made such slow progress, in comparison of the other practical arts, may be partly referred to the difficulty and intricacy of the art itself, and partly to some peculiar disadvantages which the profession lies under, and which I shall afterwards endeavour to explain.

2. There is a certain metaphysical subtlety, which is not only useless in our inquiries into nature, but does real mischief, by giving genius and industry a wrong direction. This involved all science, for many ages, in darkness and controversy. It was carried to the greatest length by the schoolmen, many of whom [161] having great acuteness, abundance of leisure, from their monastic life, little acquaintance with good authors, and still less with the works of Nature, spun out of a small quantity of matter, those cobwebs of learning, curious indeed for the fineness of the thread, but of no substance or utility. As their writings consisted of subtleties, and a play of words; as they occasioned perpetual wranglings, and led to no useful consequences, the wiser part of mankind became dis-

gusted with them, so that now the old school-philosophy has fallen into universal contempt. This philosophy corrupted no science more than medicine. From the days of Galen, till the middle of the last century, all the institutions of physic were filled not only with the doctrine of elements and temperaments, but with enquiries, Whether the procuring of health be the design or end of medicine? whether disease is a quality or relation? and such like. They were generally disputes about words; and whenever the terms [162] were defined, the controversy was at an end. It is really melancholy to reflect on the industry, erudition, and often genius, that was wasted in such disputes as disgrace the human understanding, and was employed in corrupting an art, that more requires attentive and sagacious observation, than metaphysics, to bring it to perfection.

An useless subtlety may be displayed in two ways; either in the prosecution of enquiries of no importance, but of difficult investigation, or by treating important subjects in a way that leads only to fruitless speculation and controversy. We have examples of the first in the old school-logic, and in most metaphysical disquisitions, ancient or modern. I acknowledge the usefulness of such, considered as an exercise for young minds. They may sharpen the invention, strengthen and improve the reasoning faculty, and tend to fix the attention: but when long dwelt upon, they withdraw the at-[163]tention from the study of Nature and the practical arts; they tend to make men rather ingenious disputants than solid reasoners, and beget a habit of wrangling upon every subject, extremely disagreeable in conversation. The practice of balancing things with a finical precision, is unfavourable to the enlarged views of genius, the advancement of the sciences, and the successful management of business in private life. These require only an attention to probabilities, to leading principles, and to the great outlines of objects, a quick discernment where the greatest probability of success lies, and habits of acting, in consequence of this, with facility and vigour.

We treat important subjects in a manner that leads only to fruitless speculation and controversy, when we labour in a minute discussion of supposed necessary preliminaries, and points [add 'which we fancy to be' in errata] essentially connected with them, though, in [164] reality, they have no connection at all, or a very remote one. It is the same useless labour, when we pierce beyond certain limits into Nature, and attempt the investigation of causes, either beyond our reach, or such as, if known, could lead us to no useful consequences. Thus philosophers, before Sir Isaac Newton appeared, often attempted to explain the cause of gravity.

But that great man contented himself with investigating the laws according to which it acts, and only proposed a conjecture of its cause in the modest form of a query. The laws according to which gravity, magnetism, and electricity act, are a proper subject of enquiry; because they are within our reach, and because the knowledge of them leads to the most useful consequences. But their causes will probably ever elude our deepest researches, nor, perhaps, would the discovery be useful. The reciprocal influence of the soul and body is one of the most important enquiries in medicine; but an [165] investigation into the nature of this union, is equally obscure and unnecessary.

3. There is another species of useless subtlety, which consists in a scrupulous exactness, in regard to arrangement and method. These should, without doubt, be attended to in treating of any subject, but are peculiarly necessary in all the departments of natural history. The proper distribution of plants, and other productions, into their several orders, genera, and species, is a great assistance to the memory, and leads to the knowledge of their virtues. But a complete classification is a matter of the greatest difficulty, and can never be attained, without a knowledge of all the particulars proposed to be classed. It may be attempted upon different principles, as is the case with the various systems of botany, and though one of them may, upon the whole, be more perfect than the rest, yet each of them may have its peculiar advantages. In the same man-[166; numbered 169 in error]ner, diseases may be classed according to their symptoms, their remote or proximate causes; and in various other ways, all of them very imperfect, but each useful in some degree.

There are many conveniences attending a proper arrangement of diseases. By bringing those together that have a natural affinity, the history of one disease throws light on that of another, and by comparing the circumstances in which they agree, general principles are formed in regard to the genus or order. But this advantage can only attend the arrangement of diseases upon the principle of natural and real affinity; as, for example, in intermitting fevers, topical inflammations and haemorrhages; but it cannot be obtained from arrangements purely artificial. The different orders of diseases comprehended by nosological writers under the class of *cachexiae*, do not agree in any circumstances of real connexion, from which the class can be defined. [167] Most, even of the particular genera, forming its different orders, are dissimilar in every material circumstance. A proper arrangement of diseases is of service, by making it easy for a physician, who finds difficulty in a case of practice, to compare it with similar ones related by authors. It likewise facilitates the communication of observations by shortening descriptions.

The want of clear and precise definitions has been the cause of much confusion and disputation in medicine, as well as in other branches of science, abstract mathematics excepted. It seems now to be agreed, that it is more convenient, on the whole, to define the genera of diseases by a simple enumeration of such symptoms as are most constantly present, as are obvious to the senses; and which serve to distinguish them from others which they most resemble. Definitions of diseases ought not to include any hypotheses in relation to their proxi-[168]mate causes; nor should they at all point at such hypotheses; otherwise physicians, unless their opinions of proximate causes are the same, can never agree in annexing the same ideas to the same words. When diseases are defined by a simple enumeration of obvious symptoms, there is little room left for disagreeing about the name to be given to any patient's complaints. Definitions of diseases ought, as far as possible, to be taken from symptoms existing together; but sometimes it is necessary, in order to characterise the disease, to enumerate symptoms as they occur in succession, as in the cases of intermittent and exanthematous fevers. They should not, if it can be avoided, include symptoms which happened in the beginning of the disease, of which perhaps the patient can give no account; nor should they depend upon the duration of a disease, because that is always uncertain. Sometimes it is necessary to include the external or occasional cause of the disease in the definition, [169] since the symptoms alone are not sufficient to distinguish it from others of a different nature. But the external cause should never make a part of the definition, unless it be obvious. Established names ought not to be changed without some very urgent reason; but whenever medical writers have generally agreed to give a name to a certain assemblage of symptoms, that term should, on no account, be applied to a different assemblage, to prevent confusion.

After all, it is impossible to define the genera of diseases with such accuracy, as not to leave it doubtful, sometimes, to what genera some particular cases should be referred. There is greater difficulty in giving a systematic arrangement of diseases than of bodies in natural history, arising from the frequent uncertainty of their diagnostic symptoms, from their symptoms not being permanent, and from the frequent complication of diseases with one another. [170]

They who have hitherto attempted to class diseases methodically, have differed widely from one another in regard to the number, distribution and definitions of the classes, orders and genera. What some have enumerated as genera, others have considered as species or as symptoms; nor perhaps is it possible for human ingenuity to remove the difficulties and imperfections attending every attempt of this kind, until the knowledge of particular diseases, and the science of medicine, is rendered perfect.

It is evident, therefore, that this subject of arrangement presents an ample field for disputes, by which the attention is diverted from the study of diseases themselves, and of the proper method of treating them, to fruitless speculations about the order in which they should be ranged. I call them fruitless, only so far as they waste too much of that time and attention which might be more usefully employed. Upon this subject I [171] would recommend to you Dr. Cullen's arrangement of diseases, as not only natural and simple, but on account of the clearness and precision of his definitions.

If we carry our studies in natural history no farther than to a just arrangement, and a knowledge of names, what we have learned is of no more consequence, than the knowledge of a Greek grammar, and of the words in a Greek dictionary, would be to one who was never to look into a Greek writer. I speak of natural history with regret, as I see its principal purpose too much neglected. I see it studied rather as a matter of curiosity, or as furnishing subjects of ingenious speculation, than as subservient to real utility. It is of little importance to settle the classes, orders, and genera of plants, in comparison of ascertaining their uses; yet the one subject has been attended to very closely, the other has been worse than neglected; it has been corrupted by many false facts, [172] especially in what relates to medicine. Much pains have been taken to place those worms that infest the human body in their proper ranks, and to examine their structure with the greatest accuracy; but little proportionable care has been taken to ascertain the signs of their existence in the body, the symptoms they produce there, and the most effectual method of destroying them. I cannot, however, omit, on this occasion, doing justice to the merits of the great Linnæus, who has displayed so original a genius, in reducing all the subjects of natural history into so perfect and beautiful a system. Nor has he stopt here; he has shewn the most enlarged spirit of observation, in applying natural history to the useful purposes of life, particularly to agriculture and medicine.

4. The advancement of the sciences has been much retarded by the credulity of those who have cultivated them. This credulity discovers itself by an easy acquiescence in what are asserted to be facts, [173]

although not properly authenticated; in a fond belief in the powers of certain delusive arts, in a bigotted attachment to some great names, or in a superstitious veneration for antiquity.

(a) An easiness of belief, in regard to facts, by admitting them without authority, has corrupted every branch of natural knowledge, but none of them so much as medicine. Facts depending upon the animal oeconomy, must be difficult to ascertain; because it is subjected to sudden and unexpected changes, from various causes which we cannot trace, and often not depending on any material causes, but on some unknown affection of the nervous system. A heated imagination, therefore, may easily magnify them, and fraud may easily counterfeit them, whilst, at the same time, it is difficult to detect the error. Hence our accounts of the effects of remedies still remain full of uncertainties and falsehoods; while many other branches of natural history, chemistry in particular, have of late been [174] happily cleared of them. Medicine suffers more from this cause than even from theories. The weakness of a theory is easily detected. The understanding of one sensible man is sufficient of this. But it frequently requires the united labours of many to distinguish facts that are fully and candidly represented, from such as are false or exaggerated; nor can it be done until an opportunity offers of repeating the observation or experiment, perhaps at the risque of a patient's life. I do not mean to insinuate here, that no facts should be admitted into natural history, or medicine, but such as are thoroughly established. I mean only to shew the impropriety of mixing uncertain reports and undoubted truths, without making a proper distinction between them. Whatever is asserted to be a fact, although somewhat extraordinary, and supported by slender authority, yet deserves to be recorded, till the truth of it can be ascertained; nothing shews more ignorance of Nature, or more self-sufficiency, than to [175] reject facts merely because we cannot account for them.

(b) A fond belief in the powers of certain delusive arts, particularly astrology, natural magic, and alchemy, has greatly retarded the progress of knowledge, by engrossing the attention of men of genius, and by introducing, into medicine especially, a multitude of false facts, founded on superstition and delusion. These arts, which promised to be of use in life, laid such hold on the imagination, that no power of reason was able to free men from their enchantment. At the same time, they have accidentally given rise to some curious discoveries, and their effects on the mind would furnish some excellent materials for the history of the human imagination.

(c) A bigotted attachment to certain great names, has done much mischief to science. The history of philosophy exhibits to the world, from time to time, [176] some man of distinguished ingenuity who has erected a system. This system has been adopted for a few years. Learned men have commented upon it; some have diffusely explained it; others have abridged it. In the mean time, none of those authors rose higher than their source; few of them so high. In the succession of a few years another original genius has arisen, exposed the weakness of his predecessor's system, and established another in its stead. This, after having the like honours paid to it by commentators, expositors, and epitomisers, has sunk, in its turn, into contempt and oblivion. This has been the fate of medicine, from the days of Hippocrates down to the present time, when there appears to be a general disposition to throw off the shackles of authority, to appeal to Nature in matters of fact; and to assert the right of private judgment in matters of opinion and reasoning. I do not mean to insinuate the possibility of every individual's thinking for him-[177]self in these matters. Nature never intended the bulk of mankind either to think for themselves, or to act from principles of their own. I only mean to express my regret, that men, blessed with superior talents, should be swayed by an authority they ought to have controuled, and should assent to doctrines, which a little exercise of their own judgments would have shewn to be ill-founded.

(d) Another obstacle to the improvement of science, similar to the former, has been a blind and superstitious veneration for antiquity. It is inconceivable to those who are acquainted only with the present state of the learned world, and with the free spirit of enquiry that now prevails, to what an absurd height this attachment to antiquity was formerly carried; how much it has cramped the efforts of genius, and retarded the progress of knowledge. Yet if we consider the source of this attachment at the time when it chiefly prevailed, it ap-[178]pears to have been natural and excusable. Upon the decline of the Roman empire, all the useful sciences and elegant arts decayed apace, and at last, by successive irruptions of barbarians, were entirely extinguished. A cloud of ignorance overspread mankind till towards the end of the fifteenth century. From time to time, however, some sparks of genius broke through the gloom, and fortunate accidents preserved some of the most valuable remains of ancient arts and wisdom. Medicine underwent the fate of the other sciences, and slept in the same darkness. About the middle of the fifteenth century Constantinople was taken by the Turks, and many of the Greek manuscripts found there, were brought into Italy by Theodore Gaza. The noble art of Printing was discovered about the same time, which soon spread those treasures of antiquity over Europe. About this period, so important in the annals of history, and so big with great events, men began to awake out of that [179] lethargy in which they had been so long sunk. Upon the first discovery of the Greek and Roman writers, the visible superiority of their sense, their taste, and their elegance, beyond what the world had seen for many ages, was soon perceived and acknowledged. It was therefore to be expected, that men of science and ingenuity should at that time have employed themselves in recovering, translating and commenting on the remains of antiquity which had escaped the ravages of time and barbarism, and lain for many centuries buried in the cells of monks. How much was the world obliged to those restorers of learning! The immediate effects produced by the recovery of the ancient writers, shewed clearly in what their principal excellency consisted. All the fine arts, painting, sculpture, architecture, speedily rose to a high degree of perfection. Purity of language, and an elegant simplicity of composition, especially in poetry and history, were particularly studied. But natural [180] history and natural philosophy remained much neglected. The reason was this: in all works of taste and imagination in poetry, in eloquence, in simplicity, correctness, and elegance of composition, the ancients possessed an excellence hitherto unrivalled. In abstract mathematics, likewise, they will ever remain as standards of that clearness and precision which should be the characteristics of mathematical reasoning. But in natural history, and in natural philosophy, they were not equally successful. This was owing partly to their not having bestowed sufficient attention on those subjects, and partly to this, that these sciences depend for their advancement, not so much on the genius of one man, as on the accumulated labours of many. Thus a Homer, an Appelles, a Praxiteles, or a Demosthenes, may have carried poetry, painting, sculpture, or eloquence, as high, or higher, than any who have succeeded them; because, when these men died, their arts, in a great measure, died with them. [181] But in natural history and natural philosophy, the case is widely different; because every man, who applies to any branch of these studies, may avail himself of all the labours and improvements of his predecessors. As the knowledge of Nature, then, at the revival of learning, was in a low state; and as little light was thrown on it by the writings of the ancients, they continued to lie almost neglected, till towards the middle of the last century; men of learning and ingenuity, before that time, generally devoting their attention to theological studies, the fine arts, and the different branches of polite and ancient literature.

The same warm admiration of antiquity which prevailed in other sciences at the restoration of learning, very properly attached physicians to the ancient writers in their own profession. It had been happy, however, for mankind, if, instead of a blind admiration of Hippocrates, justly styled the father and foun-[182]der of medicine, they had imbibed some portion of his spirit for observation. Hippocrates will always be held in the highest esteem, for his accurate and faithful description of diseases, for his candour, his good sense, and the simple elegance of his style. But instead of prosecuting his plan, and building on the foundation he had laid, his successors employed their time in commenting on his works. Galen began with writing largely on what he reckoned the genuine productions of Hippocrates, in which he endeavours to reconcile all his seeming contradictions, and to prove the truth of his observations by a variety of arguments, not founded on his own extensive experience, but on the Aristotelian philosophy; some of them, indeed, subtle and ingenious, but for the most part weak and sophistical. This manner of commenting on books of observations, is extremely absurd. The first enquiry here ought to be into the truth of the facts. Till these are confirmed by similar [183] observations, it is a waste of time and labour to attempt an explanation of their causes. Hippocrates has left us a number of excellent observations; [add 'together with' in errata] some that are found to be true only in certain cases, and under certain limitations; some peculiar to the climate and country in which he lived, some so obscure that they cannot be understood, some ill founded, and a great number that seem curious and important, in regard to which not one of his numerous commentators has taken the trouble to enquire, whether they were true or false. Every one of them has, after the example of Galen, attempted to prove the truth of his observations, not by similar observations of their own, but by hypothetical reasoning, drawn from the prevailing philosophy of the times they lived in. Thus the noble foundation of observations begun by Hippocrates, and the example he has set of faithful and accurate description, have, in a great measure, been neglected, while physicians, in all ages, [184] have fondly attempted to support their opposite theories by his authority, in which they were favoured by the obscurity of some parts of his writings. Not only his observations, but his opinions, (of which indeed he was very sparing) till very lately, were opposed to the authority of facts, which appealed for their truth to the experience of every man of candour and common sense; so that a physician, in writing his own observations, found himself under a sort of necessity to shew that they agreed with those of Hippocrates, at least that they did not contradict them. The effect of this was, that the truth of Nature was often perverted in order to make it correspond with the sentiments of Hippocrates, or even with the authority of Galen. This introduced a corruption into the very source of all solid knowledge in medicine; and at the same time, encouraged a pompous display of learning in writing on medical subjects, that wasted the time and tired the reader, who [185] wanted to know what Nature said, not what Hippocrates and Galen thought, in medicine. Neither is this pedantry yet extinct in Europe; there being few medical books written in some parts of it, which are not stuffed with numerous quotations from the ancients, containing some trite observations, that answer no other purpose, but to make a parade of erudition.

5. Another obstruction to the progress of science, the reverse of the former, has been a fond attachment to novelty. This proceeds partly from a principle in the human mind, which is gratified with whatever is new, independent of other considerations; partly from an anxiety to discover truth upon an interesting subject, which makes us often grasp a shadow for the substance; and partly from a disposition to believe whatever we wish to be true. The uncertainty of the methods of cure, in many diseases, makes patients, and sometimes physi-[186]cians, eagerly adopt any new method that promises a more effectual and speedy remedy. This is the source of that universal propensity to give credit to the extravagant accounts of the effects of nostrums and quack medicines. These are recommended to a patient, with an assurance of their infallibility, an assurance which no prudent or honest physician gives to any remedy in any disease whatever. From the same cause we have seen, in our own times, many remedies highly praised for their efficacy, in almost every disorder, and soon after neglected; such as, cold water, crude mercury, soap, tar-water, lime-water, sea-water, Ward's medicines, and even some of the poisons. All these, in their turn, were deemed infallible; and when time discovered the folly of this expectation, they have been with the same precipitation almost wholly laid aside; as if a medicine could not be useful in the cure of any one disease, because it was not success-[187]ful in all. This attachment, however, to novelty, is not such a bar to improvement, as a superstitious veneration for antiquity. The former, from time to time, is bringing new accessions to knowledge: the latter keeps the active powers of the mind suspended in a vain admiration of what, perhaps, was of some

value in the infancy of science, but what is now universally known. A physician of coolness and sagacity may avail himself of these temporary intoxications of the public, in regard to such remedies, as they give him an opportunity of ascertaining the effects of some, from seeing them administered in larger doses, and for a greater length of time, than patients would otherwise be persuaded to take them. The passion for novelty is indeed particularly excusable in medicine; because it is natural for us to be pleased with what seems, not only to bring an accession to our knowledge, but to communicate some useful discovery. [188]

6. The hasty reduction* [*"Bacon" at bottom p. 189]] of any science into a system, apparently full and perfect in all its parts, while, in reality, these parts are ill filled up and erroneous, is a bar to its farther improvement. The intention of these systems is to place a science in the most favourable light. It is, therefore, delivered in a magisterial manner, so as to acquire credit without examination: and hence a science descends in the persons of master and scholar, not of inventor and improver. Men are generally attached to systems; because they free them from the impatience of doubting, and promise them certain principles on which their minds may securely rest: and teachers find it contributes both to their interest and reputation, to reduce the sciences into systems, seemingly complete. One who appears well acquainted with the principles of a science, and who seems to entertain no doubt of their soundness, makes a better appearance, than one who [189] doubts, and fairly owns that he does so. The bulk of mankind are not judges of the merit of men or deep science; and are ready enough to allow to pretenders the consequence they assume, if they do not too much over-act their part. I have already endeavoured to shew the propriety of prosecuting enquiries into Nature upon a regular and methodical plan. In teaching a science, it is equally necessary to proceed upon a plan of arrangement. But, till all the facts and principles included in a science are fully established, it is impossible to reduce it into the form of a regular system; and there are many circumstances relating to arrangement, which, in the mean time, must remain undetermined. It is, therefore, sometimes better to use the unconnected aphoristical manner, than to attempt an order, or at least to be very solicitous about an order, where there are no certain principles to lead to it. [190]

It has been the fate of medicine to suffer, in a particular manner, from this rage of systemizing. It has fallen, at different times, into the hands of Galenists, Chymists, Cartesians, Mathematicians, Stahlians, and some other sects compounded of these; each of whom have moulded the whole sciences into a form, seemingly complete in all its parts. It has been tinctured with mystical divinity, astrology, and all the subtleties of school philosophy, according to the different attachments of physicians to those studies. But, notwithstanding the disadvantages attending these systems, a physician of genius will be able to draw from them some useful information.

7. The last impediment* [*"Bacon" at bottom p. 191] I shall mention to the progress of science in general, has been, too great attention to purity and elegance of language, on the one hand; and, on the other, an affected obscurity and intricacy of style. In works [191] of taste, and addresses to the passions, a language highly ornamented may be very proper; elegance, sublimity, pathos, are there in their proper place. But the language in which science is to be communicated, should be simple, perspicuous, and divested of all artificial ornaments. Original writers, who have new ideas to communicate, are often obliged to use new words and phrases, the better to convey their meaning; which surely they, and they only, have a right to do, provided they clearly define them. An affected intricacy of style is now, in a great measure, laid aside. The use of technical terms, where others equally clear and expressive can be found, is regarded as pedantry, or a cloak to conceal ignorance. This censure may sometimes be carried too far, but in general it is just. That learned jargon, which so long disgraced philosophy, was introduced from a principle of vanity, or for the unworthy purposes of excluding from science, all who were not of profes-[192]sion. But it evidently retards the advancement of science, when men attend more to words than to things, whether it be in an affected display of learning, or in a scrupulous regard to purity of expression or elegance of composition.

Let me take this opportunity of recommending to your serious study the writings of Lord Bacon, who of all men possessed, perhaps, the most enlarged and penetrating genius. He has explained the method of acquiring knowledge, and promoting science, with incomparable judgment and perspicuity. He has likewise left us some beautiful specimens of true philosophical induction, particularly in his History of the Winds. This, and some other of his essays in natural history, are to be considered in no other light, than as specimens of his method of carrying on enquiries into nature. The facts they contain are not to be depended on: he was obliged to take such as were then generally received, which, whether [193] true or false, equally served his purpose. He uses a language peculiar to himself: It has been censured, and perhaps justly, with being too figurative, which renders it in some places obscure; but in general it is well fitted for communicating science, being clear, noble and expressive. [194]

LECTURE VI

Peculiar disadvantages under which medicine has laboured. – Inconveniencies attending the method in which it has been usually taught, entirely from the lectures of professors, or from books. – The advantages of a regular attendance on the sick, during the whole time in which a physician is studying his profession, particularly specified. – Duties of a professor of medicine.– Inconveniencies arising from the absolute confinement of the study and practice of physic to a class of men who live by it as a profession. – Advantages of laying the art open, and of gentlemen of science and abilities, who are not of the profession, studying it as an interesting branch of philosophy. – Attempt to shew that this would tend to promote the interests of humanity, by diffusing the [195] benefits of the art; that it would facilitate the improvement of medicine; that it would most effectually support the dignity of the profession, and secure the success of every individual belonging to it, in proportion to his real merit. – Conclusion.

I Endeavoured, in a former lecture, to explain some of the principal causes that have obstructed the progress of science in general; and, where it was necessary, applied my observations particularly to physic. I thought it necessary to explain to you my general sentiments, in relation to the improvement of knowledge; because it gave me an opportunity of communicating my leading principles in the science of medicine. But before I conclude the subject, a regard to truth and candour obliges me to take notice of some peculiar disadvantages attending medicine, and which seem to have retarded its progress. This I do, not from a desire to expose the weakness of a profession, [196] the honour of which my inclination, and many ties, lead me to support; but merely with a view to establish this honour upon a liberal and solid foundation; and to put you on your guard against certain errors and inconveniencies, to which you might otherwise be exposed. As I have the misfortune to differ from many of my brethren on this subject, while I propose my own sentiments with freedom, I wish to do it with a proper deference to their judgment. The peculiar causes which have retarded the progress of medicine, besides the difficulty and intricacy of the art, formerly mentioned, appear to me to have arisen partly from the manner in which it has been usually taught, and partly from its having been confined to a set of men who live by it as a profession.

In the first place* [*"Bacon" at bottom p. 197]] it must be observed, that the general method of conducting education in universities, does not seem [197] so well calculated to advance science as to diffuse it; not so well fitted to promote particular arts, as to communicate general principles. Those who teach the sciences, often make use of various allurements with their students; sometimes with the laudable view of engaging their attention; sometimes from a desire to give a dignity to their own characters, by pretensions to discoveries, by the triumph of confutation; the ostentation of learning, or the veil of mystery. For the conveniency of teaching medicine, it has been usual to adopt the synthetic plan; that is, to lay down general principles. especially such as relate to the proximate causes of disease, and the mode of operation of remedies, and to mention facts so far only as they serve to illustrate those principles, or as they are clearly deducible from them. Medicine, likewise, as usually taught in universities, instead of being represented as an art imperfect in its most material branches, instead of having its deficiencies pointed [198] out, with a view to their being supplied, is digested into a regular and seemingly perfect system. In this light it is beheld by the student, who embraces hypotheses with the same facility and unsuspecting confidence as he would do facts established on the testimony of his senses; he imagines he understands the causes of all diseases, and the manner in which remedies act in removing them; his mind is at ease in having always firm principles to rest on. But how fallacious these principles have generally been, is sufficiently evinced in the history of medicine, which shews that they have been constantly fluctuating. For example, a morbid acrimony of the blood has been assigned as the proximate cause of certain diseases: the nature of this acrimony has been specified: the manner in which occasional causes have produced it, has been explained; plausible indications of cure have been laid down in consequence of these supposed discoveries: remedies have been prescribed correspondent [199] to these indications; and their operation in destroying the acrimony has been pointed out. Yet perhaps, upon farther enquiry, it may be found that there is no evidence of any acrimony existing in these diseases; or that, supposing there is reason to suspect in general that there may be such a disorder in the blood, we are still ignorant of its specific nature; that we do not know in what manner the external causes produce the symptoms, whether by first vitiating the blood, or by acting immediately on the nervous system; that, from our uncertainty in regard to these circumstances, the indications of cure become likewise uncertain; that there is no proof of the remedies acting in the manner which had been supposed; and that, perhaps, some of those remedies, though in repute for many ages in the cure of such diseases, have either no effect at all, or at least none in the doses commonly given. In short, it may sometimes be found, that all we certainly know of the matter, is, that certain ex-[200]ternal causes produce these diseases; that experience has ascertained the effects of some remedies in curing them; and that this experience is the only rational basis on which we can ground our future practice.

A student, however, is seldom aware of the fallacious nature of such hypothetical structures, as he is a stranger to the circumstances on which they are founded. They appear plausible, well connected, and are particularly grateful, as they tend to conceal the difficulties of the profession.

Medicine has little chance of acquiring improvement from a physician educated in the faith of systems, because he scarcely supposes it admits of any. He treats his patients according to the established rules, and when they die, he is satisfied that every thing was done for them that art could do. It might be supposed that enlarged experience, and [201] the riper exercise of his understanding, would remove his prejudices; but a little acquaintance with mankind shews, that early and strong impressions are with great difficulty erazed; every circumstance that tends to confirm them, is readily attended to, while every one that tends to weaken them, is overlooked or ingeniously explained away; so that time seems frequently to confirm our errors.

It is, indeed, difficult and painful for men to give up favourite opinions, and to sink from a state of security and confidence into one of suspence and scepticism. Accordingly we find that physicians do not easily change the principles they first set out with. We have remarkable examples of some, who, after having written systems of medicine early in life, have lived to be old, have been admired for their genius, have had extensive practice, and though their systems in the mean time had gone through many editions, yet no material alteration has been made [202] in them; a proof how much they were attached to their first ideas.

Although the principles of the medical art are originally established by investigation, or induction from particular facts, yet it would be both tedious and unpleasant to teach it entirely upon this plan. I am therefore of opinion, that the best method of teaching it, is to unite the synthetic method, which is most commodious for communicating knowledge, with the analytic one, which leads to improvements and inventions. If medical education was conducted in this manner, a student would be, in some degree, an eye-witness of the observations and experiments upon which the principles of the science are founded. For example, if he were daily conversant among the sick, he would enjoy many advantages beyond what can be derived from books or lectures. Some of these I shall mention. [203]

1. Whatever one sees, makes a deeper and more lasting impression on his mind, than what he learns from description.

2. There are many circumstances relating to diseases and remedies, of which it is difficult to convey a just idea, viz. different appearances of the countenance, state of the pulse, breathing, voice, smells, tastes, and different degrees of heat, &c. Hence every experienced physician, or indeed artist of any profession, knows much more than he is able to communicate.

3. Diseases are described in systems as existing by themselves; but in practice they are found complicated in such various forms as no description can specify, and to which no general practical rules can be applied.

4. Medical facts are often related imperfectly; sometimes from the author's inattention to the concomitant circum-[204]stances, sometimes from his thinking them of no importance. But the truth is, facts are seldom mentioned in systems, unless with a view to establish a theory, or to recommend a medicine; and whatever facts are either not subservient to these views, or are repugnant to them, are often but slightly mentioned, or suppressed. Medical writings likewise abound with false or exaggerated accounts of the effects of particular remedies, occasioned by avarice, vanity, credulity, a warm imagination, or a weak judgement.

5. A student educated in this manner, acquires the habit of attention and discrimination; he brings the truth of general principles to the test of experience; he discovers the falsity of some of them, and learns to ascertain the many exceptions and limitations to which others are subjected; he often finds the most plausible indications of cure to be delusive, and that among the numerous remedies recommended in consequence [205] of these indications, none are able to relieve the patient. By these means he acquires an early distrust of all theories, however specious.

6. He ascertains the importance of the several branches of medicine, and of all medical enquiries, as relative to the main end of his profession, the preventing or curing of diseases; and regulates his application to them

accordingly.

7. He becomes familiarised to the humours and weaknesses of patients; he acquires some address in managing their tempers, and in soothing their distresses; a conduct which in certain circumstances is of real consequence.

8. He begins to acquire an address in the management of the sick, a quickness of apprehension, a composure and presence of mind, and some decision and resolution in sudden emergencies. A young physician, who has drawn his knowledge only from books or lectures, [206] although he may be ingenious and learned, and consequently able to talk plausibly, will yet be extremely embarrassed when he enters upon practice– Medicine is not merely a speculative science to be acquired by study alone; it is an active and practical art, the proper exercise of which can be attained only by long practice. This is allowed to be the case in all the other practical arts, and the education in them is conducted accordingly. Let us suppose [add 'of' in errata] a young man designed to be a sailor, that for the first years of his education he studies mathematics, natural philosophy, and navigation, but has never been at sea: when he makes his first appearance there, what must be his situation? He can talk of mechanical powers, of friction, of the nature of magnetical effluvia, of the theory of the winds, and, in short, shew himself master of every branch of his profession, so far as speculation could carry him. But can he handle a rope? can he go aloft and furl the sails? can he make an observa-[207]tion in a rolling sea? can he do any one useful work aboard the ship, or direct the sailors how to navigate her in a storm? Who would trust himself to the direction of such a sea commander?– The case is much the same with a young physician, who has had what is called a regular education, and is well grounded in every branch of his profession, except the practice; in which he must be defective, if he has not for some years diligently attended the sick. A careless and irregular attendance on an hospital for a few months before he settles in business, is very insufficient to qualify him for the important charge. I own, however, that a young man cannot reap much advantage by attending the sick, till he is acquainted with the rudiments of physic: But there is no impropriety in his studying and applying to practice at the same time; on the contrary, it is attended with the advantages above mentioned: and the shortness of the time usually allotted for medical education, does not allow of their being separated. [208]

9. A physician who has been educated upon this plan, whose mind has

never been enslaved by systems, because he has been a daily witness of their insufficiency, instead of being assuming and dogmatical, becomes modest and diffident. When his patient dies, he secretly laments his own ignorance of the proper means of having saved him, and is little apt to ascribe his death to his disease being incurable. There are indeed so few diseases which can be pronounced in their own nature, desperate, that I should wish you to annex no other idea to the word, but that of a disease which you do not know how to cure. How many patients have been dismissed from hospitals as incurables, who have afterwards recovered, sometimes by the efforts of unassisted nature, sometimes by very simple remedies, and now and then by the random prescriptions of ignorant quacks? To pronounce diseases incurable,* [*"Bacon" at bottom p. 209] is to establish indolence and [209] inattention, as it were, by law, and to skreen ignorance from reproach. This diffidence of our own knowledge, and just sense of the present imperfect state of our art, ought to incite us to improve it, not only from a love of the art itself, but from a principle of humanity. I acknowledge, however, that such a diffidence as I have described, if it be not united with fortitude of mind, may render a physician timid and unsteady in his practice; but, though true philosophy leads to diffidence and caution in forming principles, yet, when there is occasion to act, it shews how necessary it is to have a quickness in perceiving where lies the greatest probability of success; to be decisive in forming a resolution, and firm in carrying it into execution.

It is much in the power of a teacher of the art, to obviate the inconveniences commonly chargable upon systems. It is his duty, in treating of any subject, to give a full detail of facts; to separate [210] real from pretended ones; and to arrange them in such a manner as may lead to the discovery of causes and general principles. If these cannot be established by a just induction, he may, with propriety, suggest an hypotheses; but, while he gives his reasons for its probability, he should, with equal impartiality, state every objection against it. So far from throwing a veil over the numberless imperfections of his art, he should be solicitous to point them out, and at the same time direct to such observations and experiments as may tend to remove them. Sensible of the warm imagination and credulity of youth, of their proneness to admiration, and their eagerness to have every fact accounted for, he ought to guard against the errors into which these dispositions may lead them, and should endeavour to direct their ardour in the pursuit of knowledge to proper subjects; not to those that merely amuse the fancy, but to such as exercise the powers of useful observation and invention; to [211] subjects of real and permanent importance.

I throw out these observations with freedom, because I am well acquainted with the liberal spirit that prevails in this university, in every department of science, and in none more than in all the branches of medicine. To this university I owe, in a great measure, my own education; but there are none of my obligations to it which I remember with more gratitude, than the acquisition of some portion of that freedom of spirit, for which it has been always distinguished. The medical societies of students which have been conducted with decency and regularity, have in this, as well as in other respects, produced the best effects. In these, they have been taught to feel and exercise their own powers; to arrange their ideas, and to express them with facility; and that honourable emulation has been excited, which is a principal spring of diligence and activity. Let me take this [212] opportunity of doing justice to the merit of several gentlemen, who have, within these few years, done honour to this medical college, by their inaugural dissertations. In these, several important investigations have been carried on, by a set of accurate and well-conducted experiments, under the direction of my learned and ingenious colleagues. This method of giving a specimen of a young physician's genius, is attended with so many advantages, is so creditable to himself, and so useful to the public, that I should be extremely sorry to see it fall again into disuse. – But to return.

I shall now endeavour to shew, that the confinement of the study and practice of physic, entirely to a class of men who live by it as a profession, is unfavourable to the progress of the art.

Nothing can so effectually tend to the improvement of an art, as the making it the interest of those who practice it, to [213] contribute to its improvement. But it happens unfortunately, that the spirit and application required to the advancement of medicine, is often checked by a necessary attention to private interest. Physicians are influenced by the same general motives of action with other men. Some of them love medicine, and would gladly devote their time and attention to it, so far as their situation could admit; others practice it merely as a trade. But the state of our profession is singular. A common artificer has no other way of rendering himself eminent in his trade, but by excelling in it. Of this, all mankind are judges. If he be a bad workman, no address or qualifications of any other kind can avail him. No gentleman can hope to rise in the profession of the law, who does not possess the abilities of a lawyer. The proofs of his knowledge, ingenuity and eloquence, are daily exhibited to the world, and their value is duly ascertained. The public have the same opportunity of estimating [214] the merits of a divine. In short, every man's merit, in his profession, may be well known to the public; and is in general suitably rewarded. The science of medicine alone is kept so carefully concealed from the world, and the art must necessarily be practised in so private a manner, as renders it difficult for the public to form a just estimate of a physician's knowledge from the success of his practice. Accordingly, in no other profession is the reward of merit so uncertain. If a physician be only acquainted with the outlines of practice, and have a good share of address and common sense, he may succeed well. This success is not surprising, if he be generally esteemed as a man of genius and knowledge in other subjects; because, it is presumed, that these will extend to his own profession. But it is more unaccountable, though the case frequently occurs, to see physicians rising to great eminence, who, far from possessing learning or abilities of any kind, are known to be men of weak [215] understandings. – Those people seem to have a strange idea of physic, who trust their lives in the hands of a man, whose discernment and common sense they would despise on any other occasion.

The check which the improvement of medicine must receive, from withholding the reward justly due to those who excel in it, is sufficiently obvious – A physician, when he sets out in the world, soon perceives that the knowledge most profitable for him, is not that merely of his profession. What he finds more essential, are the various arts of insinuation and ostentation. This leads to views very different from those of genius and science. To his real merit as a physician, he cannot easily find a patron, because none are judges of this but those of his own profession, whose interest it often is to have it concealed.

By what I have said, I mean only to represent the disadvantages naturally con-[216]sequent on leaving it to physicians to judge of the merits of their brethren. It is putting human virtue to too severe a trial, and indeed it is a trespass against the most obvious maxims of prudence and humanity, to suffer people to be tried by judges whose interest it is to condemn them. Nor do I mean, in making an observation which is equally applicable to every class of men, to include all the individuals in our profession. There is a virtue found among many of them, which can stand the severest test, and there is an elevation of mind, that generally accompanies genius, which renders those who possess it equally superior to the suggestions of envy or interest, and to all the low arts of dissimulation. The difficulties which regular physicians encounter in attempting to introduce any improvement in practice, is one of the principal causes which induce such of them as love the science, to turn their attention to some other branch of medicine, which they can cultivate with more safety and [217] freedom. In all these, their discoveries have been numerous and useful. But how rapid a progress would the practical part of medicine make, if physicians were at equal liberty to improve it, under the inspection and patronage of men qualified to judge of their merit, and who were under no temptation, from sinister motives, to depreciate it [substitute 'their conduct' for 'it' in errata]?

It were to be wished, that ingenious men would devote half the time to the study of nature, which they usually give to that of opinions. If a gentleman has a turn for observation, the natural history of his own species is surely a more interesting subject, and affords a larger scope for the display of genius, than any other branch of natural history. If such men were to claim their right of enquiry into a subject that so nearly concerns them, the good effects in regard to medicine would soon appear. They would have no interest separate from that of the art. They would detect and expose [218] assuming ignorance, and would be the judges and patrons of modest merit. Cases often occur, where a physician sees his patient hastening to dissolution, he knows a remedy that affords some prospect of saving his life; but it is not agreeable to common practice, and is dangerous in its operation. Here is an unhappy dilemma. If he gives the remedy, and the patient dies, he may be ruined; for his conduct will be watched with a malignant eye. But if the scheme of gentlemen of fortune applying to the study of physic should take place, the encouragement and assured protection of knowing and disinterested judges, would animate a physician in his practice. Such judges, not fettered by early prejudices, unawed by authority, and unbiassed by interest, would canvass with freedom all the commonly received principles of medicine, and expose the uncertainty of many of those maxims of which a physician dares not seem to doubt. [219]

There are some advantages, which gentlemen who study medicine only as an interesting branch of natural philosophy, possess, beyond physicians by profession. A physician, amidst the necessary duties and anxieties of an extensive practice, has little leisure to attend to any subject that is not directly connected with his business; nor does he always possess that tranquility of mind which is so requisite in every kind of investigation, and particularly in planning and conducting a train of experiments. Lord

Bacon had as enlarged views in medicine, of its deficiencies, and of the proper method of supplying them, as perhaps any physician who ever wrote. Dr. Hales has been one of its greatest benefactors, by his discoveries; by the openings he made in different branches of the science, which have since been further prosecuted, but principally by the excellent example he set of ingenuous and accurate experimental investigation. Cornaro, a Venetian nobleman, when some [220] years turned of fourscore, composed a little treatise on regimen, written with singular candor, simplicity, and precision. With more pleasure could I name Mr. Boyle on this occasion, had not his credulity lessened that esteem, which his diligence, genius, and many virtues, so well merited.

But, not to insist further on arguments to shew, that there is less reason to expect improvements in our profession, while it continues on its present narrow footing, I shall only observe, that it appears from the history of physic, that the improvements in the practical parts of it, have seldom been owing to those who valued themselves upon being regular, systematic, rational practitioners; nay, what is more extraordinary, such improvements have been often opposed by them with keenness and acrimony, and seldom adopted till after a long struggle. I could give instances of this in the case of blisters, opiates, Peruvian bark, antimony, [221] mercury, inoculation of the small-pox, and I may add, the cool regimen in fevers. Many important discoveries relating to the cure of diseases, have been made by accident: and some valuable remedies have been communicated to us by the natives of America, and other illiterate nations. But, till of late, it would be difficult to point out many solid improvements in practice which have been the result of reasoning, or of any regular train of observations or experiments. On the contrary, the merits of [delete 'the merits of' in errata] the improvements that have been offered to the world, and which appealed for their justness to experience, have been usually adopted, not upon repeated and more accurate trials, but upon the authority of great names, or from the prevailing philosophy of the times.

There has been much reason to complain, that the discoveries of those men, who were not of the medical faculty, have not been always examined with that [222] candor, which their importance and success required. Yet from such men very substantial improvements may sometimes be expected. Even quacks possess some advantages in their practice beyond regular physicians, as they seldom can suffer much, either in their interest or reputation, from the bad success of their experiments. But they have another advantage above the regular physician, from having more extensive practice. I grant, however, that the ignorance and inattention of most of these men, makes them profit but little, in proportion to what might be expected, from their experience, and unfettered practice; and I own too, that little regard can be had to their veracity, in their accounts of cures. But it is a physician's duty, to search for knowledge from all sources, however impure and contemptible; and he may avail himself of that experience, which the empiric himself is neither able nor willing to turn to account. It was from strolling chymists and the lowest artificers, [223] and not from the schools of philosophy, that Mr. Boyle drew that large and useful collection of facts with which he has enriched many branches of science. I must, however, observe, with pleasure, that the same freedom of enquiry, which has enlightened every other branch of natural knowledge, begins now to extend to medicine; that the tyranny of authority and systems declines apace; and that there is a fair prospect of the science being rebuilt on the more solid basis of nature, on facts, and an accurate induction from them.

It is said by those who insist on the propriety of confining the study of physic to a class of men who live by it as a profession, That the science is so abstruse, that it cannot be understood but by a person who devotes himself entirely to that study. The little progress it has made, notwithstanding the labours of so many ingenious and learned men wholly directed to its cultivation, is brought as [224] a proof of its difficulty. It is said, that if people were encouraged to study physic who are not regularly bred to it, and who do not intend to follow it as a profession, quacks would be multiplied, and patients would lose that confidence in the physician, which is as necessary for their own sakes as for his. It is further said, that a smattering of physic could tend only to fill people's minds with imaginary diseases, and apprehensions of danger upon slight indisposition.

These reasons have appeared so powerful to many of the faculty, that they have watched with a jealous eye over all intruders, and have often treated them with abuse and ridicule, even when it was apparent that they were actuated purely by motives of humanity. It would not be candid to ascribe this to any sordid views: Enlarged knowledge produces a liberal and unsuspicious spirit: and no profession can boast of more men of learning, ingenuity, and liberal education, [225] than ours. But as the reasons, above assigned, for the absolute confinement of the study of medicine to physicians, do not appear to me satisfactory, I shall take the liberty of examining them particularly.

The difficulties, which a gentleman, not intended for the profession, is to encounter, in acquiring some share of medical knowledge, are greatly exaggerated. Some of them are real and unavoidable; but the greater number are either imaginary, or arise from the mysterious form in which the science lies concealed, unnecessarily involved in technical terms, and incumbered with inquiries of no utility, or not applicable to practice. Medicine, in one point of view, is a science of boundless extent; but this should not deter any person from the study of it, as the same might be said of every branch of natural knowledge. In our prosecution of any of them, the farther we advance, the more sensible we [226] become of their difficulty, and of the further improvement of which they are capable. The argument, however, brought to shew the difficulty and intricacy of the medical art from the slowness of its improvement, (notwithstanding the joint labours of so many physicians employed in this single pursuit) may be obviated, by observing, that if by medicine be meant, the art of preserving health and curing diseases, the truth is, that very few physicians of genius have endeavoured to cultivate it, and that some of those few have attempted it in a way that could not reasonably be expected to succeed.

It will be readily owned that a physician who has regularly studied the several parts of medicine, must possess great advantages even in regard to practice, above a gentleman, who has only attended to them in a more cursory manner. But their is no reason to say, that one must be perfect master of these parts, before he can attain such a knowledge of the [227] practice, as may be in some degree useful, when the assistance of an able physician cannot be procured. Surely it is not a matter of such difficulty, for a gentleman of a liberal education, to learn so much of medicine as may enable him to understand the best books on the subject, and to judge of the merits of those physicians to whom he commits the charge of his own health, and the health of those more immediately under his care and protection. It is difficult to ascertain to what extent a gentleman should be instructed in medicine, before he can pretend to practice as above mentioned. The most that can be required of him, is, such a degree of knowledge as is commonly possessed by practitioners of acknowledged merit, and such knowledge as physicians, educated in different schools of medicine, and attached to different theories, concur in judging essential. – In this view, it is evident, that he should know as much anatomy as is necessary to understand the animal oeco-[228]nomy in its sound and morbid state; that he should know the principles of chymistry, particularly in their application to pharmacy and the other parts of medicine; that he should be acquainted with the history of diseases, especially with those circumstances that serve to distinguish one from another, when apparently similar, but really of a different nature, and requiring a different method of cure; and that he should be instructed in the nature of the usual remedies. The knowledge of these last-mentioned branches, should be drawn from medical writers of distinguished sagacity, accuracy, and candour; but above all, from observation and experience, the purest and least fallible source of medical science. He may derive singular advantages from the conversation of an ingenious and experienced physician, who is able to direct his studies, to distinguish between genuine and pretended facts, and, amidst the load of learned rubbish with which medicine is encumbered, to select what [229] is truly useful. Such a course of study as I have described, though somewhat formidable at first view, is really not so to those who love science, and who have laid a tolerable foundation of learning. There are, indeed, difficult cases that often occur in practice, which require the assistance of the greatest medical discernment; but any man of good understanding may comprehend the general principles of the theory and practice of physic, if the facts, on which they are founded, are full and clearly laid before him.

The objection to laying medicine open to the world, like other sciences, from its tendency to multiply quacks, and to lessen the authority of the physician, is not well founded. It is not possible to confine the practice entirely to regular physicians. Cases are continually occurring of people labouring under diseases, who can have no access to the assistance of one of the faculty. It would be bar-[230]barous to hinder those from using such remedies as appeared to them most likely to afford them relief, or to prohibit a friend or a bystander from giving assistance in such a situation. In fact, as every person prescribes occasionally, the only question is, whether they should receive any assistance from art, or be left to act as their fancy may lead them. If, by withholding this assistance, every disease, where a physician was not consulted, was to be left to nature alone, physicians would have a plausible excuse for keeping the world in ignorance; because it might be alleged, that more diseases would be cured by the efforts of unassisted nature, than by the random management of people imperfectly instructed in medicine. But, in reality, this is never the case in diseases of any consequence. I shall give an example, in the general treatment of fevers among the lower class of people, when they are deprived of medical assistance.– The unhappy patients are gene-

[231]rally confined to a close room, where they breathe a hot and a putrid air; every method is tried to raise a sweat; they are loaded with bedclothes; sometimes they are made to drink spiced and strong liquors; at other times large quantities of warm water gruel, although their stomach loathe it, and it occasion flatulence, sickness, and oppression. If, in consequence of great heat or delirium, they attempt to get out of bed, they are confined to it by force; nor are they suffered to change their bed- or body-linen, till the fever is quite removed; by which means the air becoming more putrid, aggravates the symptoms, and makes the disease contagious. – In such cases, because the patients have no physician, and take no medicine, the disease is said to be left to nature. But this is a mistake. If such patients had been really left to nature, they would have been treated very differently. They would have been indulged in whatever was agreeable to them; they would have breathed cool [232] and fresh air; they would not have been teazed to eat or drink beyond what their appetite demanded; they would have been indulged with cold water or small beer in what quantity they pleased; they would have been suffered to get out of bed and to enjoy the cold air, or to have had few bedclothes, with liberty to throw out their limbs without controul; their linen would have been changed daily, and every thing kept clean and sweet about them. Similar instances might be produced from other diseases. Patients are so far from being left to nature, when no physician is called, that they are commonly oppressed with a succession of infallible cures recommended by quacks, or by their weak and officious friends.

I must here observe, that there is a suspicion entertained against physicians, as rejecting all remedies proposed by those who do not belong to the faculty, especially if their composition be kept a secret. Whatever the case may have [233] been formerly, or may still be among a few individuals, the censure is now ill founded. Every remedy which has the appearance of usefulness, meets with a fair trial from the gentlemen of our profession. I speak this with the more confidence in regard to those of the British dominions, where medicine is in general practised with much candour and humanity; but it would be an imputation on their knowledge, and indeed on their common sense, if they were to give credit to all the accounts of cures which daily impose on the credulity of mankind.

Physicians, in their early practice, are sometimes controuled and intimitated from doing what they think necessary for the recovery of their patients; not by people whose education and knowledge should make their opinions respectable, but by the most ignorant, and consequently, most conceited, part of mankind. Physicians have nothing to fear from the intrusion of men of science [234] who have turned their attention to medicine. Such will be modest in proportion to their knowledge of the subject, and will be the readiest to call for the assistance of a physician of experience and abilities, to respect his judgment, and to enforce his prescriptions; whilst, at the same time, himself [substitute 'they' for 'himself' in errata] may suggest what may be useful, to the ablest of the profession.

If we consider the situation of a young physician of genius, brought forward and supported in his profession under the honourable patronage of those who are judges of that genius; and that of another, destitute of such assistance, and compelled by necessity to attend to the prejudices, and to humour the caprices of the ignorant and impertinent intruders into his office; how pleasant, how creditable is the one? how humiliating the other, to every man of spirit and sensibility? [235]

I have thus endeavoured to shew that, by laying medicine open, and encouraging men of science and abilities, who do not belong to the profession, to study it, the interests of humanity would be promoted, the science would be advanced, its dignity more effectually supported, and success more certainly secured to every individual, in proportion to his real merit.

Before I conclude, I must observe, that such objections as are made against any person pretending to judge of medical subjects, who has not been regularly bred to the profession, were formerly urged against the reformers from Popery. Besides the divine authority claimed by the church, it was said, that a set of men, who devoted their whole time and studies to so deep and complicated a subject as theology, were the only proper judges of whatever belonged to it; that calling their authority in question was hurting [236] the cause of religion, and lowering the sacerdotal character. Yet experience has shewn, that since the Laity have asserted their right of enquiry into these subjects, theology, considered as a science, has been improved; the real interests of religion have been promoted; and the clergy have become a more learned, a more useful, and even a more respectable body of men, than they ever were in the days of their greatest power and splendor.

I hope I have advanced no opinions in these Lectures that tend to lessen the dignity of a profession which has always been considered as most honourable and important. But, I apprehend, this dignity is not to be supported by a narrow, selfish, corporation-spirit; by self-importance; by a formality in dress and manners, or by an affectation of mystery. The true dignity of physic is to be maintained by the superior learning and abilities of [237] those who profess it, by the liberal manners of gentlemen; and by that openness and candour, which disdain all artifice, which invite to a free enquiry, and thus boldly bid defiance to all that illiberal ridicule and abuse, to which medicine has been so much and so long exposed.

FINIS. [238]

NOTES

1. Thomas Sydenham (1624-1689), a British physician, is one of the most important figures in the history of 17th-century medicine. He contributed to the clinical science of fevers, to the effects of weather on disease, and to nosology.
2. Richard Mead (1673-1754) was, measured by his income, an immensely successful practitioner. He wrote on quarantine as a measure to control plague.
3. Samuel Garth (1661-1719) wrote a famous poem, 'The Dispensary'. In this poem he attacks the greed of apothecaries for opposing the founding of a dispensary for the poor.
4. Joshua Ward (1685-1761), a well known quack of the 18th century, made his fortune from the sale of medicines, including a powder for dropsy. In *Biblioteca Osleriana* Osler describes Ward as "perhaps the most successful and the best of the 18th-century quacks."
5. Georg Ernest Stahl (1660-1734) understood health and disease in terms of a vital power of the body. He aimed for an explanation of disease in terms of general processes and his work was approved of by Gregory.
6. Carolus Linnaeus (1707-1778) made major contributions to nosology, including the classification of plants. He also developed a classification scheme for diseases into eleven major classes, divided in Aristotelian fashion into genera and species.
7. The Reverend Stephen Hales (1677-1761) made various contributions to the ventilation of ships and jails, as a preventive medicine measure. He also made contributions to the measurement of blood pressure in animals and to the understanding of the mechanics of circulation of blood.
8. Luigi Cornaro (c.1463-c.1566) wrote on hygiene in his 80's, blending an eclectic, personal approach with a religious approach in his account of diet and longevity. He also called for the physician to be the friend of the patient.
9. This is probably Themison, a second-century Roman physician, usually understood to be the founder of the "Methodist" school. He was a student of Asclepiades of Bithynia.
10. Paracelsus (c.1493-1541) contributed to nosology. He held that diseases were caused by things independent of the person who became sick (the ontological concept of disease). These agents invaded the body, disturbing its physiology and thus causing diseases.
11. William Harvey (1578-1657) relied on observation rather than dogmas and developed a crucial account of the circulation of the blood.
12. Giorgio Baglivi (1648-1701) called for rigorous bedside observations of diseases and their patterns.
13. Lorenzo Bellini (1643-1704) contributed to renal anatomy, including the identification of the straight tubules and the renal excretory duct (Bellini's duct).
14. Fabricus Hildanus (1560-1624) was a German surgeon, credited with being the first to recommend amputation of gangrenous limbs above the involved tissue.
15. Marcus Aurelius Servinus (1580-1656) wrote what some regard as the first textbook of surgical pathology, part of the "new anatomy" following Vesalius (1514-1564).
16. Thomas Bartholin (1616-1680) approached anatomy as a mathematical and mechanical science.
17. John Arbuthnot (1667-1735) undertook experimental investigations of the air and its role in epidemic diseases.

18. Hermann Boerhaave (1668-1738) developed mechanistic accounts of anatomy and physiology and became the dominant figure at the medical school in Leiden. Although he had died before Gregory went to Leiden for medical studies, Boerhaave's influence remained very strong there and at the University of Edinburgh, where his texts were taught in the medical curriculum by his students (Underwood, 1977).
19. Friedrich Hoffmann (1660-1742) wrote an important treatise on medical ethics, *Medicus Politicus* (1749), which bases the role of the physician on Christian ethics and the prudential calculation of self-interest. Gregory reacts against the latter sort of approach to medical ethics, arguing, instead, for a sympathy-based, moral life of service to patients. Hoffmann was also an advocate of the concept of a "vital principle" to explain animal physiology.
20. Thomas Reid (1710-1796) was Gregory's cousin and fellow member of the Aberdeen Philosophical Society and a major philosophical figure of Scottish moral sense and Enlightenment philosophy.

INDEX

Aberdeen Grammar School, 7
Aberdeen Philosophical Society, 6, 10-13, 15, 36, 40, 59, 248 n. 20
  Baconian commitments of, 11
  and Hume, 11-13
  Questions of, addressed by Gregory, 11, 59-66
  on skepticism, 11
  on sympathy, 11
animalicula, 25
Arbuthnot, John, 116, 185, 247 n. 17
Aristotle, 209
Asculapius, 89
Aspin, Richard, 67
Athanasius, St., 12-13

Bacon, Francis, 1, 24ff, 82, 95-96, 127, 150, 158, 199, 201, 204, 218, 228-31, 239
  on geriatrics, 28
  on offices of medicine, 28-29
  on "outward euthanasia," 29
  on prolongation of life, 28
Baglivi, Giorgio, 91, 247 n. 12
Baker, Robert , 2
Bartholin, Thomas, 110, 178, 247 n. 16
Beaton, James, 71, 89
Beattie, James, 11
Bellini, Lorenzo, 91, 247 n. 13
bioethics
  ahistoricism of, 2-3
  invention of by Gregory, 1-3, 42-44
Black, Joseph, 16
Blagden, Sir Charles, 67
Blair, Hugh, 12
Bluestocking Circle, 10
Boerhaave, Hermann, 7-8, 24ff, 43, 64, 81, 91, 116, 134, 178, 185, 206, 247 n. 18
  influence of, 248 n. 18
Bowman, Inci, 93, 161
Boyle, Robert 127, 155, 199, 239

Campbell, George, 11-12
Carlyle, Alexander, 8
Chalmers, Principal, 7
Comparative View of the State and Faculties of Man with those of the Animal World, A, 5, 11, 15
Cornaro, Luigi, 83, 158, 239, 247 n. 8
Crawford, James, 30-31
Cullen, William, 4, 15-16, 67, 153, 22

Death
  definition of, 41-42
  determination of, 41-42
Descartes, René, 25, 136, 215
Digby, Sir Kenhelm, 31-33
Dunbar, James, 11

Eccles, Audrey, 20-21
Elements of the Practice of Physic, 16
enthusiasm, 13
Essay on Genius, 11
Essay on the Nature and Immutability of Truth, An, 11
Essays on the History of Mankind ..., 11
ethics
  feminine, 2
  feminist, 2
etiquette, 2
euthanasia
  outward, 29
  perfect, 17
  regarding preparation of the soul, 29

False manners, man of, 20-21
Father's Legacy to his Daughters, A, 9-10, 16
fiduciary, 42
Forbes, Lord William, 9
formality of dress, 69-70
Foster, J., 71

Galen, 90, 140, 146-47, 209, 226
Garth, Samuel, 62, 247 n. 3
Gaub, Jerome, 32
Gaza, Theodore, 225
Gerard, Alexander, 11-12
Gevitz, Norman, 26-27
Gordon, Thomas, 7

Gregorie, James, 7
Gregory, Elisabeth, 9, 36-38
 death of, 13
 Gregory's portrait of, 14-15
Gregory, James, 5-6, 17
Gregory, Janet, 7
Gregory, John
 and anticipation of his own death, 16
 Baconian commitments of, 26-32
 biographical sketch of, 6-17
 on competition among practitioners, 63, 153-59, 236-39
 contributions of, to history of medical ethics, 1-3
 death of, 16-17
 on death of his wife, 13-15
 on drinking, 68-69
 on existence of God, 116-18, 127-28, 186, 199
 fellowship of in Royal College of Physicians of Edinburgh, 15
 and feminine medical ethics, 36-39
 as a feminist, 36-39
 as First Physician to his Majesty the King in Scotland, 16
 on gout, 16
 honors accorded to, 8
 on humanity, 98-99
 on improvement of our nature, 28
 in London, 9-10
 mathematical ability of, 7
 on metaphysics, 81
 on the mind-body relation, 128, 200
 as Professor of Medicine at the University of Aberdeen, 10
 as Professor of Medicine at the University of Edinburgh, 15-17
 as Professor of Philosophy at King's College, 8-9
 on quacks, 83
 Questions of, to Aberdeen Philosophical Society, 59-66
 relationship of, to Hume, 12-13
 ridicule of, for his views on women, 38
 on ridicule of physicians, 72, 97, 164-65
 on scientific method, 126-59, 198-230
 on virtue, 81
 on women, 8
 on women of learning and virtue, 36-39
Gregory's medical ethics, 17-44
 on aging physician, 63, 73, 172, 177
 on animal experimentation, 78, 190
 on atheism, 64-65, 70, 76-77, 87, 116-18, 127-28, 186, 199-200
 background of, 40
 and bioethics, 32
 on care of dying patients, 22, 87, 175
 on clergy's role in care of dying, 22, 175
 as clinical, 43
 on confidentiality, 68, 74, 87, 104-105, 172-73
 on conflict of interest, 63, 153-57, 167, 216-17, 236-40, 242-43
 on consultation, 76, 86, 109-10, 175-78
 on decorum, 69-70, 76-77, 86-87, 107, 113-14, 174
 on the definition of medicine, 28, 71, 96, 128, 164, 200
 on delicacy, 77, 114, 181-83
 development of, 3-6
 on diffidence, 40, 42
 on dignity of the medical profession, 97-98, 165-66, 244-45
 on emergencies, 102, 169-70
 on experiments on patients, 216-19
 on fees, 69, 104, 108-109
 as feminine, 1-2, 43, 102-104
 on formality of dress, 113, 180-81
 on genius of the physician, 63-64, 67-68, 73, 81, 98, 100-102, 111, 123, 166, 168-69
 on governance of the patient, 68-69, 74, 86, 103-104, 107, 171, 174
 on hardness of heart, 98, 118
 on humanity, 68, 72, 85, 98-99, 102-104, 108, 118, 166-67, 170-72
 on infidelity of the physician, 116-19, 185-87
 on laying medicine open, 156-59, 236-45
 and Macbeth, 103
 on manners of the physician, 69-70, 76-77, 86-87, 107, 113-14, 182
 on medical corporations, 176-77
 on medical degrees, 112, 176, 180
 on nostrums and secrets, 70, 77, 82, 87, 115-16, 147-48, 183-85
 on obligations of a professor of medicine, 235-36
 on openness to conviction, 75, 86, 106-107, 173-74, 239-40

on paternalism, 107
on patients of High Rank, 70, 73-74, 103-104, 114-15, 170-71, 183
on patients of low rank, 70, 73-74, 103, 114-15, 170-71, 183
on patients with nervous diseases, 22-23, 68, 74, 104, 171-72
as philosophical, 41-42
on physician as fiduciary, 42
on physician as friend, 102
on the physician-patient relationship, 102-104
on the physician's responsibility when a patient dies, 62-63, 83, 152, 234-35
on the practice of medicine, 164
on prejudices, 101-102
on prescribing, 75
as professional medical ethics, 1, 42-43
on quack medicines, 83, 147-48
on qualifications of a physician, 40, 100-102, 168-71
on regard for older writers, 83-84, 125, 146-47, 177-78, 196, 226
on relationship of physicians to apothecaries, 69, 76
on relationship of physicians to surgeons, 69, 76, 110-12, 178-80
on religion, 8, 64-65, 70, 76-77, 87, 116-19, 185-86, 199-200
on seduction of female patients, 68, 74-75, 104-105
on servility of physicians to social superiors, 70, 114-15, 183
on steadiness, 39-40, 42, 102, 169-70
student notesets of, 4
on surgery, 76, 87
on sympathy, 2, 30-32, 36-39, 40, 43, 72-74, 85, 102-105, 165-66, 170-72
on temperance of the physician, 106, 173
on tenderness, 36-40, 42, 86, 102-105, 170-71
on time management, 183
topics of, 40-42
on truthtelling, 69, 75-76, 80, 87, 106-108, 174-75
on utility of medicine, 59-66, 72, 97, 164
on vanity, 117
on virtues of the physician, 40, 99, 102-104
on visits to the patient, 104, 183
on women of learning and virtue as moral exemplars, 36-39, 43
on young physicians, 63, 73-74, 99, 101, 105, 117, 167-69, 172, 177, 243-44
Gregory's philosophy of medicine, 24-32
on anatomy, 120-21
on art of medicine, 59-66
on authority in clinical judgment, 82, 100-101
on botany, 122
on capacities of medicine, 28, 71, 96, 128, 164, 200
on chemistry, 65, 78
on clinical judgment, 134-35, 206-207
on clinical reasoning, 206-208, 210-14, 241-42
on concept of disease, 60, 82-84, 91-92, 111, 178-79, 220-22
on connections among the branches of Physic, 119-23, 188
on diagnosis, 91-92
on dignity of medical art, 97-98, 159
on the Dogmatical Physician, 213-14
on the empirics, 210-13
on French-language medical texts, 79, 125, 196
on Greek-language medical texts, 79, 125, 196
on history of medicine, 89-91, 193-94, 207-14
on improvement of medicine, 64, 233-40, 242-44
on incurable disease, 62-63, 83, 152
on Latin-language medical texts, 79, 124-25, 195-96
on laying medicine open, 156-59, 236-45
on *materia medica*, 122, 190
on mathematics in medicine, 194-95
on mechanical principles of the body, 79, 120, 189
on medical books, 197-98
on medical education, 65, 72, 77, 83, 98, 119-26, 152-53, 165-66, 187-98, 231-35
on medicine, 61
on obstacles to the improvement of medicine, 150-59, 230-33, 236, 240
on pathology, 79, 84
on pharmacy, 65-66
on physiology, 78-79, 119-22, 188-91

on practice of medicine, 192
on self-interest of physicians, 156-59, 236-37, 240
on systems in medicine, 232
on teaching medical science, 61-62
on Theory of Physic, 191, 213-14
on therapeutics, 84, 122, 190-91
on utility of medicine, 59-66, 72
on vernacular-language medical texts, 125-26, 196-97
Gregory's philosophy of science, 24-32
on analogies, 132, 204-205
on branches of science, 80, 128, 200-201
on causal principle in reasoning, 129, 201-202
on diffidence, 63, 132-33, 205-206
on the end of science, 139-40
on existence of God, 127, 199-200
on experience, 129-35, 201-14
on importance of Bacon, 82, 150, 199-200, 229-30
on intellectual temperaments, 136-38
on laws of nature, 80, 135-36, 204
on natural history, 81, 137-39, 217
on nature, 60, 199, 126, 129, 131, 135-36, 188, 198
on obstacles to improvement of science, 132-34, 139-50, 204-206, 208-10, 218-30
on public utility of science, 127-28, 199-200
on role of creative imagination in science, 135
on excessive subtlety in science, 140-43, 218-20
on system-building in science, 135, 148-49, 221-22

Hales, Stephen, 83, 158, 239, 247 n. 7
Haller, Albrecht von, 41
Harvey, William, 90-91, 116, 185, 247 n. 11
Hildanus, Fabricus, 110, 178, 247 n. 14
Hippocrates, 89-90, 125, 135, 144, 146-47, 178, 194, 207-208, 226-27
Hoffmann, Friederich, 22, 42, 81, 116, 134, 185, 206, 248 n. 19
on animal spirit, 25
Horace, 95
Hume, David, 1-2
Baconian commitments of, 36
on sympathy, 32-36
on understanding

Inquiry into the Human Mind ..., An, 11
is-ought derivation

King's College, Aberdeen, 7
honorary degrees from, 8
medical school of, 10
regent system of, 7

Lawrence, Christopher, 7
Lawrence, Paul, 6
Lectures on the Duties and Qualifications of a Physician, 4-6, 16, 40-44
errata in, 163
Leeuwenhoek, Anton von, 25
Linnaeus, Carolus, 81, 143, 247 n. 6
Lyttleton, Lord George, 5

Macbeth, 39, 103, 170
Lady, 39
Macleod, I.B., 85
man-midwives, 20-21
collusion of, in birth of illegitimate children, 21-22
manners, 19-20
crisis of, 20-22
Mead, Richard, 62, 247 n. 2
medical corporations, 94-95
medical ethics
ahistorisism of, 2-3
and etiquette, 2
Gregory's invention of, 1-3
history of, 2-3
medicine, Baconian, 26-32
medicine, eighteenth-century, 9, 17-24
competition in, 17-18, 20
concept of disease in, 17
explanation in, 18
manners in, 19-20
obstetrics in, 20-22
openness of, 18
physician-patient relationship in, 18-24
power of patient in, 18-19
self-treatment in, 18
trust in, 19, 22
Medicus Politicus, 22
midwives, 17-18
Monro, Alexander, *primus*, 7, 153

Montagu, Elizabeth, 10, 15, 36-38
  Queen of the Bluestocking Circle, 10
  as role model for Gregory, 36-38
moral strangers, 34

Newton, Sir Isaac, 81, 127, 136, 141, 199, 215

Observations on the Duties and Offices of a Physician ..., 4-6, 16, 40-44
obstetrics, 20

Paracelsus, 90-91, 110, 154, 178, 247 n. 10
Peruvian bark, 64, 134, 206, 239
Philosopher's stone, 90
Philosophy of Rhetoric, 11
physician-patient relationship, 18-24
  as fiduciary relationship, 42
  Gregory's invention of, 42-43
Physiological Library, 32, 36
physiology , 29
Plato, 208-209
Poker Club, 38
Porter, Dorothy, 17ff
Porter, Roy, 17ff
President's Commission for the Study of Ethical Problems in Medicine and Biomedical and Behavioral Research, 42
Pringle, Sir John, 162
profession, Gregory's invention of medical, 42-43
"Proposall for a Medicall Society, A," 27-28

Rather, L.J. , 31
Reid, Thomas, 7, 10-13, 34, 129, 202, 215, 248 n. 20
Risse, Guenther, 4, 18, 23-24
Robinson, Elizabeth. *See* Montagu, Elizabeth
Royal Infirmary of Edinburgh, 3-4, 23-24
  admission to, 23
  patients' experience of, 23-24
  patrons' role in, 23
  physician-patient relationship in, 23
  scarcity of resources in, 23
  as teaching institution, 23
  treatment in, 23-24
Rutherford, John, 4, 7-8, 15

Science, Baconian, 24-32
  and diffidence, 26
  experimental method of, 24ff
  and God, 26
  metaphysical commitments of, 25
  principles in , 25
  skepticism in, 26
Scottish Enlightenment, 10-11, 42
Serapion, 207, 210
Servinus, Marcus Aurelius, 110, 178, 247 n. 15
Shakespeare, William, 38-39, 103, 170
Skene, David, 10
Smellie, William, 6, 21
social principle, 33
Stahl, Georg, 81, 91, 116, 134, 185, 206, 247 n. 5
  on vital principle, 247 n. 5
steadiness, 36-39
  Gregory on, 36-40, 42, 86, 102-105, 169-71
Stevenson, A.M., 85
Stewart, Agnes Grainger, 5-6, 9
Sydenham, Thomas, 60, 64, 116, 178, 185, 206, 247 n. 1
sympathy, 30-32
  as action at a distance, 33
  as consent, 30-31
  and consensus, 31
  as contagion of imagination, 31-32, 35
  Crawford on, 30-32
  Digby on, 31
  as double relation of impressions and ideas, 33-36
  false, 38
  feminine, 36-39
  Gaub on, 32
  gendered, 36-39
  Gregory's account of, 2, 40, 43, 72-74, 85, 102-104, 170-72
  Hoffmann on, 32
  Hume on, 32-36
  and *idiopathia*, 30
  and milk of human kindness, 38-39, 103, 170
  and moral strangers, 34
  as physiologic principle, 32, 35-36
  as principle of human nature, 32-33, 35-36
  and *protopathia*, 30
  real, 103, 171
  and social principle, 33

and steadiness, 39
and surgery, 34

Telescope, reflecting, 7
tenderness, 36-39
Gregory on 36-40, 42, 86, 102-105, 170-71
Terrence, 34
Themison, 90, 247 n. 9
Thicknesse, Philip, 21
Tong, Rosemary, 2
Treatise of Human Nature, 11-13, 32ff
Truman, John, 6-7
Tytler, Alexander Fraser, 6, 16

Ulman, H. Lewis, 10-11
University of Edinburgh, tradition of *lehrenfreiheit* in, 236

Ward, Joshua, 64, 147, 154-55, 227, 247 n. 4
Whytt, Robert, 16, 22-23

Young, Thomas, 4

Philosophy and Medicine

1. H. Tristram Engelhardt, Jr. and S.F. Spicker (eds.): *Evaluation and Explanation in the Biomedical Sciences*. 1975 ISBN 90-277-0553-4
2. S.F. Spicker and H. Tristram Engelhardt, Jr. (eds.): *Philosophical Dimensions of the Neuro-Medical Sciences*. 1976 ISBN 90-277-0672-7
3. S.F. Spicker and H. Tristram Engelhardt, Jr. (eds.): *Philosophical Medical Ethics*. Its Nature and Significance. 1977 ISBN 90-277-0772-3
4. H. Tristram Engelhardt, Jr. and S.F. Spicker (eds.): *Mental Health*. Philosophical Perspectives. 1978 ISBN 90-277-0828-2
5. B.A. Brody and H. Tristram Engelhardt, Jr. (eds.): *Mental Illness*. Law and Public Policy. 1980 ISBN 90-277-1057-0
6. H. Tristram Engelhardt, Jr., S.F. Spicker and B. Towers (eds.): *Clinical Judgment*. A Critical Appraisal. 1979 ISBN 90-277-0952-1
7. S.F. Spicker (ed.): *Organism, Medicine, and Metaphysics*. Essays in Honor of Hans Jonas on His 75th Birthday. 1978 ISBN 90-277-0823-1
8. E.E. Shelp (ed.): *Justice and Health Care*. 1981 ISBN 90-277-1207-7; Pb 90-277-1251-4
9. S.F. Spicker, J.M. Healey, Jr. and H. Tristram Engelhardt, Jr. (eds.): *The Law-Medicine Relation*. A Philosophical Exploration. 1981 ISBN 90-277-1217-4
10. W.B. Bondeson, H. Tristram Engelhardt, Jr., S.F. Spicker and J.M. White, Jr. (eds.): *New Knowledge in the Biomedical Sciences*. Some Moral Implications of Its Acquisition, Possession, and Use. 1982 ISBN 90-277-1319-7
11. E.E. Shelp (ed.): *Beneficence and Health Care*. 1982 ISBN 90-277-1377-4
12. G.J. Agich (ed.): *Responsibility in Health Care*. 1982 ISBN 90-277-1417-7
13. W.B. Bondeson, H. Tristram Engelhardt, Jr., S.F. Spicker and D.H. Winship: *Abortion and the Status of the Fetus*. 2nd printing, 1984 ISBN 90-277-1493-2
14. E.E. Shelp (ed.): *The Clinical Encounter*. The Moral Fabric of the Patient-Physician Relationship. 1983 ISBN 90-277-1593-9
15. L. Kopelman and J.C. Moskop (eds.): *Ethics and Mental Retardation*. 1984 ISBN 90-277-1630-7
16. L. Nordenfelt and B.I.B. Lindahl (eds.): *Health, Disease, and Causal Explanations in Medicine*. 1984 ISBN 90-277-1660-9
17. E.E. Shelp (ed.): *Virtue and Medicine*. Explorations in the Character of Medicine. 1985 ISBN 90-277-1808-3
18. P. Carrick: *Medical Ethics in Antiquity*. Philosophical Perspectives on Abortion and Euthanasia. 1985 ISBN 90-277-1825-3; Pb 90-277-1915-2
19. J.C. Moskop and L. Kopelman (eds.): *Ethics and Critical Care Medicine*. 1985 ISBN 90-277-1820-2
20. E.E. Shelp (ed.): *Theology and Bioethics*. Exploring the Foundations and Frontiers. 1985 ISBN 90-277-1857-1
21. G.J. Agich and C.E. Begley (eds.): *The Price of Health*. 1986 ISBN 90-277-2285-4
22. E.E. Shelp (ed.): *Sexuality and Medicine*. Vol. I: Conceptual Roots. 1987 ISBN 90-277-2290-0; Pb 90-277-2386-9

23. E.E. Shelp (ed.): *Sexuality and Medicine.* Vol. II: Ethical Viewpoints in Transition. 1987 ISBN 1-55608-013-1; Pb 1-55608-016-6
24. R.C. McMillan, H. Tristram Engelhardt, Jr., and S.F. Spicker (eds.): *Euthanasia and the Newborn.* Conflicts Regarding Saving Lives. 1987 ISBN 90-277-2299-4; Pb 1-55608-039-5
25. S.F. Spicker, S.R. Ingman and I.R. Lawson (eds.): *Ethical Dimensions of Geriatric Care.* Value Conflicts for the 21th Century. 1987 ISBN 1-55608-027-1
26. L. Nordenfelt: *On the Nature of Health.* An Action-Theoretic Approach. 2nd, rev. ed. 1995 ISBN 0-7923-3369-1; Pb 0-7923-3470-1
27. S.F. Spicker, W.B. Bondeson and H. Tristram Engelhardt, Jr. (eds.): *The Contraceptive Ethos.* Reproductive Rights and Responsibilities. 1987 ISBN 1-55608-035-2
28. S.F. Spicker, I. Alon, A. de Vries and H. Tristram Engelhardt, Jr. (eds.): *The Use of Human Beings in Research.* With Special Reference to Clinical Trials. 1988 ISBN 1-55608-043-3
29. N.M.P. King, L.R. Churchill and A.W. Cross (eds.): *The Physician as Captain of the Ship.* A Critical Reappraisal. 1988 ISBN 1-55608-044-1
30. H.-M. Sass and R.U. Massey (eds.): *Health Care Systems.* Moral Conflicts in European and American Public Policy. 1988 ISBN 1-55608-045-X
31. R.M. Zaner (ed.): *Death: Beyond Whole-Brain Criteria.* 1988 ISBN 1-55608-053-0
32. B.A. Brody (ed.): *Moral Theory and Moral Judgments in Medical Ethics.* 1988 ISBN 1-55608-060-3
33. L.M. Kopelman and J.C. Moskop (eds.): *Children and Health Care.* Moral and Social Issues. 1989 ISBN 1-55608-078-6
34. E.D. Pellegrino, J.P. Langan and J. Collins Harvey (eds.): *Catholic Perspectives on Medical Morals.* Foundational Issues. 1989 ISBN 1-55608-083-2
35. B.A. Brody (ed.): *Suicide and Euthanasia.* Historical and Contemporary Themes. 1989 ISBN 0-7923-0106-4
36. H.A.M.J. ten Have, G.K. Kimsma and S.F. Spicker (eds.): *The Growth of Medical Knowledge.* 1990 ISBN 0-7923-0736-4
37. I. Löwy (ed.): *The Polish School of Philosophy of Medicine.* From Tytus Chałubiński (1820–1889) to Ludwik Fleck (1896–1961). 1990 ISBN 0-7923-0958-8
38. T.J. Bole III and W.B. Bondeson: *Rights to Health Care.* 1991 ISBN 0-7923-1137-X
39. M.A.G. Cutter and E.E. Shelp (eds.): *Competency.* A Study of Informal Competency Determinations in Primary Care. 1991 ISBN 0-7923-1304-6
40. J.L. Peset and D. Gracia (eds.): *The Ethics of Diagnosis.* 1992 ISBN 0-7923-1544-8

Philosophy and Medicine

41. K.W. Wildes, S.J., F. Abel, S.J. and J.C. Harvey (eds.): *Birth, Suffering, and Death*. Catholic Perspectives at the Edges of Life. 1992 [CSiB-1] ISBN 0-7923-1547-2; Pb 0-7923-2545-1
42. S.K. Toombs: *The Meaning of Illness*. A Phenomenological Account of the Different Perspectives of Physician and Patient. 1992 ISBN 0-7923-1570-7; Pb 0-7923-2443-9
43. D. Leder (ed.): *The Body in Medical Thought and Practice*. 1992 ISBN 0-7923-1657-6
44. C. Delkeskamp-Hayes and M.A.G. Cutter (eds.): *Science, Technology, and the Art of Medicine*. European-American Dialogues. 1993 ISBN 0-7923-1869-2
45. R. Baker, D. Porter and R. Porter (eds.): *The Codification of Medical Morality*. Historical and Philosophical Studies of the Formalization of Western Medical Morality in the 18th and 19th Centuries, Volume One: Medical Ethics and Etiquette in the 18th Century. 1993 ISBN 0-7923-1921-4
46. K. Bayertz (ed.): *The Concept of Moral Consensus*. The Case of Technological Interventions in Human Reproduction. 1994 ISBN 0-7923-2615-6
47. L. Nordenfelt (ed.): *Concepts and Measurement of Quality of Life in Health Care*. 1994 [ESiP-1] ISBN 0-7923-2824-8
48. R. Baker and M.A. Strosberg (eds.) with the assistance of J. Bynum: *Legislating Medical Ethics*. A Study of the New York State Do-Not-Resuscitate Law. 1995 ISBN 0-7923-2995-3
49. R. Baker (ed.): *The Codification of Medical Morality*. Historical and Philosophical Studies of the Formalization of Western Morality in the 18th and 19th Centuries, Volume Two: Anglo-American Medical Ethics and Medical Jurisprudence in the 19th Century. 1995 ISBN 0-7923-3528-7; Pb 0-7923-3529-5
50. R.A. Carson and C.R. Burns (eds.): *Philosophy of Medicine and Bioethics*. A Twenty-Year Retrospective and Critical Appraisal. 1997 ISBN 0-7923-3545-7
51. K.W. Wildes, S.J. (ed.): *Critical Choices and Critical Care*. Catholic Perspectives on Allocating Resources in Intensive Care Medicine. 1995 [CSiB-2] ISBN 0-7923-3382-9
52. K. Bayertz (ed.): *Sanctity of Life and Human Dignity*. 1996 ISBN 0-7923-3739-5
53. Kevin Wm. Wildes, S.J. (ed.): *Infertility: A Crossroad of Faith, Medicine, and Technology*. 1996 ISBN 0-7923-4061-2
54. Kazumasa Hoshino (ed.): *Japanese and Western Bioethics*. Studies in Moral Diversity. 1996 ISBN 0-7923-4112-0
55. E. Agius and S. Busuttil (eds.): *Germ-Line Intervention and our Responsibilities to Future Generations*. 1998 ISBN 0-7923-4828-1
56. L.B. McCullough: *John Gregory and the Invention of Professional Medical Ethics and the Professional Medical Ethics and the Profession of Medicine*. 1998 ISBN 0-7923-4917-2
57. L.B. McCullough: *John Gregory's Writing on Medical Ethics and Philosophy of Medicine*. 1998 [CiME-1] ISBN 0-7923-5000-6

Philosophy and Medicine

58. H.A.M.J. ten Have and H.-M. Sass (eds.): *Consensus Formation in Healthcare Ethics.* 1998 [ESiP-2] ISBN 0-7923-4944-X

KLUWER ACADEMIC PUBLISHERS – DORDRECHT / BOSTON / LONDON

MIX
Papier aus verantwortungsvollen Quellen
Paper from responsible sources
FSC® C105338

If you have any concerns about our products,
you can contact us on
ProductSafety@springernature.com

In case Publisher is established outside the EU,
the EU authorized representative is:
Springer Nature Customer Service Center GmbH
Europaplatz 3, 69115 Heidelberg, Germany

Printed by Libri Plureos GmbH
in Hamburg, Germany